AFTER SUSTAINABILITY

Timely but discomforting...examines the myth of progress and finds it wanting. It will help climate change deniers understand the issues. More importantly, it will help environmentalists, most of whom deny their own failure, to shift from a morally bankrupt optimism to a more realistic hope – with learning at its heart.

William Scott, *Professor Emeritus, University of Bath, UK and President of the UK National Association for Environmental Education*

An understanding of the bind we are in, why 'sustainability' has failed to get us out of it, and what an honest alternative might be is long overdue. This book looks beyond false hope and strained optimism to what that future might look like. Necessary and important.

Paul Kingsnorth, *poet, author and Director of the* Dark Mountain *project*

Human societies continue largely to ignore the increasingly dire warnings coming from climate and other environmental scientists. Whether or not you agree with this book's conclusions, it will make you think about the challenges and the scale of societal response that they demand. Such thought is currently in worryingly short supply.

Paul Ekins, *Professor of Resources and Environmental Policy and Director, Institute for Sustainable Resources, University College London, UK*

Dangerous climate change is coming.

Some people still deny that it is happening. Others refuse to recognise that it is now too late to prevent it. But both these reactions spring from the same source: our pathological attachment to 'progress', of which *sustainability* has been one more version.

After Sustainability traces that attachment to its roots in the ways we make sense of ourselves. Original and accessible, this is philosophy on the edge, written for anyone who glimpses our environmental tragedy and cares about our future.

Does the challenge to stop pretending offer our only remaining chance? Read this book and make up your own mind.

John Foster is a freelance writer and an associate philosophy teacher in the Department of Politics, Philosophy and Religion at Lancaster University.

AFTER SUSTAINABILITY

Denial, Hope, Retrieval

John Foster

Routledge
Taylor & Francis Group

LONDON AND NEW YORK

earthscan
from Routledge

First published 2015
by Routledge
2 Park Square, Milton Park, Abingdon, Oxon OX14 4RN

and by Routledge
711 Third Avenue, New York, NY 10017

Routledge is an imprint of the Taylor & Francis Group, an informa business

British Library Cataloguing-in-Publication Data
A catalogue record for this book is available from the British Library

Library of Congress Cataloging-in-Publication Data
A catalog record for this book is available from the Library of Congress

ISBN: 978-0-415-70639-1 (hbk)
ISBN: 978-0-415-70640-7 (pbk)
ISBN: 978-1-315-88857-6 (ebk)

Typeset in Bembo
by GreenGate Publishing Services, Tonbridge, Kent

Printed and bound in Great Britain by
TJ International Ltd, Padstow, Cornwall

For Clare, Paul and Richard

CONTENTS

PREFACE

The Prologue explains what this book is about, and I do not want to anticipate that explanation here, except to say a word about the title.

The book follows on from my earlier foray into these matters, *The
Sustainability Mirage*.[1] There I offered a detailed critique of the sustainability
paradigm that currently dominates our understanding of the environmental
crisis, and particularly our understanding of climate change. I still stand by
that critique. It is briefly reprised here in Chapter 1, and crops up now and
then in subsequent chapters. But I have come to recognise that what I made
of it in the rest of that book exemplified a form of denial which I seek to
diagnose and discuss in this one.

It would only be appropriate to pursue such diagnosis in public, of
course, if my errors there were symptomatic of something more generally
significant – as I believe to be the case. What I couldn't see, then, was how
anything *after sustainability* could be thinkable. Therefore, if the dominant
idea of sustainability was flawed, both as an interpretive paradigm and as
a policy tool, it had to be re-thought and re-cast, rather than abandoned.
That the whole concept could have failed us was inadmissible. Did it not
represent, after all, our only viable future?

My offer at recasting the concept need not now be recalled: suffice to say that
it was misdirected. The move beyond it that led to this present work involved
seeing that one must indeed bring oneself to ask the inadmissible question:
after sustainability, what? Not, that is, what comes after we have achieved it,
but what comes after we have recognised it as an irretrievably misconceived
framework and a delusive policy goal? And having raised that question, one
must at least try, with the best of one's given resources, to answer it.

But if the last book was wrong, why read this one? That is a fair enough challenge: the only proper response is to acknowledge it. The standard format of any book such as this – a contents page, chapters arguing in sequence, a conclusion, front and back covers – always embodies a kind of misrepresentation. Every such book is radically incomplete, and thus in its own way wrong; but one can only write it while one hasn't yet seen how and where. This book is offered in that spirit.

A related point is that the book should ideally stand or fall not by the authority of the many people on whose writings I have drawn, but by the force, or otherwise, of the case which I make with the thoughts that they have prompted. Similarly, there may be recognised, by those familiar with them, some very long-standing philosophical issues behind my claims; but these issues are of concern here not for themselves, but as I take them to help us understand the decisive *agon* of our time. I have therefore deliberately minimised the distracting presence of academic apparatus in the text. I hope readers will be able to ignore, at least on first reading, even the little superscript numbers that remain, but those irresistibly interested in sources and background will find them documented in the endnotes to each chapter, with links on to the Bibliography as necessary.

★ ★ ★ ★ ★

In working on this book, I have again benefited enormously from my association with Lancaster University, and from the dialogue with many intelligent and thoughtful people thus facilitated. Here might follow a longish list of names, but I prefer on the whole not to suggest complicity. I must in simple justice thank those colleagues and friends, both within and outside academia, who have read and commented on various parts of the text in various forms: Sam Clark, Brian Garvey, Emily Holman, Neil Manson, Fiona McLeod, Bill Scott and Garrath Williams. But absolutely no complicity is implied there either. I have ploughed a fairly lonely furrow in this work, and all its faults (together with any virtues) are my responsibility.

For all that, as with all my previous writing, it owes profoundly to the patience and support of my wife Rose.

The book is dedicated to our daughter and sons. I wish with all my heart that I could have written them something more comfortable, and that my generation could have left them a less grievous world. But the truth, too, is worth having. Only through it can real hope be reached.

Note

1 Foster (2008).

PROLOGUE: THE END OF PRETENDING?

It is only late that one musters the courage for what one really knows.[1]

(Friedrich Nietzsche)

There is a growing sense of crisis among environmentalists – and not just of the long crisis in human–ecological relations. Acute awareness of that always-worsening situation is now counterpointed by a gathering recognition of deep crisis within environmentalism itself.

This should worry everyone. If you are an environmentalist, you will already be worried. If you aren't quite one yet, you should be doubly worried, since things are not going well with the movement you worry about not supporting. And if you want nothing to do with environmentalism then, given the way the world is going, you should be *really* worried. (You probably are, or else why are you even dipping into this book?) We are all involved.

Because we are all involved, the book's argument is not addressed only to the environmentally concerned. Wanting to state clearly at the outset to whom it *is* addressed, I found that I could only say: to anyone seriously concerned about the human condition at the present time. The argument starts, however, with the environmental movement, and from the inner crisis of environmentalism, which is of a peculiar kind: it has been brought on by the inadmissible recognition of essential failure at the heart of success.

What success? anyone familiar with the movement might well ask. But this reaction is apt to mislead. There has, indeed, been dramatic change over the last few decades, both in the visibility of the issues and in the availability of things to do (at least potentially) about them. Environmental language, and to an extent some associated practices, are now common across our institutions, in Britain and increasingly throughout the world. This is strikingly different from how things were when I myself first got involved in these issues, half a lifetime ago. As an account of the dangers that civilisation faces, environmentalism has forced itself over this period – that is, with historically incomparable swiftness – from the fringes to the centre of our globalised political culture. It has certainly succeeded in changing the general state of awareness.

At the same time it is more and more apparent that, mainstreamed as sustainability or sustainable development, environmentalism has failed to reduce, even remotely adequately, the impact of humans on the biosphere. At this level it has not just failed to change the world, it has failed even to *engage* the world in the genuine pursuit of serious change. In particular, the futility of our collective efforts to do anything about human-induced global warming, with massively disruptive and dangerous climate change coming in its train, is there for all to see who will look at what is happening. The world talks the talk, but walks only as much of the walk as will enable it to go on talking.

Past the tipping point

A very simple argument makes the scale of our failure absolutely clear. This argument takes the form of a vicious little syllogism with four premises and a conclusion. We shall need to label this for future reference, so let's just call it the Vicious Syllogism. It goes as follows:

> *Premise 1*: If we do not keep average atmospheric temperature rise below 2 °C above pre-industrial levels, we are in for dangerous, unpredictable and potentially catastrophic climate change.

> *Premise 2*: If the world does not keep further anthropogenic emissions of CO_2 equivalent to no more than (say) 1,300 billion tonnes, we shall not keep average atmospheric temperature rise below 2 °C.

> *Premise 3*: If we are not *now* even minimally embarked on a programme that might make limiting ourselves to such a carbon budget even remotely feasible, we shall not keep further anthropogenic emissions of CO_2 equivalent to no more than 1,300 billion tonnes.

> *Premise 4*: We are not now even minimally embarked on such a programme.

So (by Premises 4 back through 1):

Conclusion: We are (already) in for dangerous, unpredictable and potentially catastrophic climate change.

Like all logic, the Vicious Syllogism is about not being able to have it both ways. The relevant science now supports the truth of the first two hypothetical premises, as environmental campaigners have persistently pointed out – they are well-founded in both attested observation and (ultimately) the laws of physics. Premise 3 seems to be equally well grounded in sober political realism. But the truth of the categorical Premise 4, that we aren't doing remotely enough by way of emissions reduction to have any chance of hitting the necessary target, should also by now be glaringly obvious. Since the argument itself is valid, the conclusion simply follows, and environmentalism, which has all along been very much about listening to the science and facing up to the facts, should in consistency stop pretending otherwise.

Key to Premise 1 here is the critical role of tipping-points in environmental change, and (again) especially in relation to global warming. The figure of a 2°C global average temperature increase over pre-industrial levels is so important because scientists have identified it as the level above which positive feedback starts to speed up the hitherto slow incremental process of warming. This positive feedback involves the earth's vegetation itself starting to emit, rather than absorbing, CO_2, thus adding a huge burst of acceleration to the greenhouse process. Thus an increase of greater than 2°C is likely to generate a runaway warming trend up towards a 6°C increase. The science writer Mark Lynas outlines the mechanisms of this progression starkly in his book *Six Degrees* – a scenario that he convincingly characterises as 'hell on earth':[2]

> If … we cross the 'tipping point' of Amazonian collapse and soil carbon release which lies somewhere above two degrees, then another 250 parts per million of CO_2 will unavoidably pour into the atmosphere, yielding another 1.5°C of warming and taking us straight into the four-degree world. Once we arrive there, the accelerated release of carbon and methane from thawing Siberian permafrost will add even more greenhouse gas to the atmosphere, driving yet more warming, and likely pushing us on into the five-degree world. At this level of warming … oceanic methane hydrate release becomes a serious possibility, catapulting us into the ultimate mass extinction apocalypse of six degrees.

It is true that Lynas is a science journalist, and many actual scientists tend to be chary about using the tipping-point argument in policy contexts. The Intergovernmental Panel on Climate Change (IPCC), for instance, sedulously avoids it, relying instead on elaborating a range of scenarios for the consequences of different increases in global average temperatures. In part this reflects a natural reluctance to give predictive hostages where all the numbers are at best indicative rather than precise. It also acknowledges a degree of uncertainty about the physical mechanisms: 'tipping point' suggests a sudden decisive shift in balance, when what may actually occur is more like a gradual slide, an increase in downward momentum that is found after a while to have become unstoppable. But the crucial idea of a *point of no return* is implicit in either model, and indeed in the very notion of linear increases in pressure on a system whose absorptive capacities are finite, and which must therefore eventually start feeding back self-damagingly. And there is a broad consensus around a 2 °C increase as the best practical guess at this point of no return within the global climate system.

So climate change campaigners have very properly emphasised the significance of this deadly threshold – but, inevitably, in a way that now hoists them firmly on their own petard. Every new authoritative report[3] provides evidence for the truth of Premise 2. In the run-up to the Copenhagen Summit of 2009, targets of 40 per cent cuts in emissions for the developed world were taken as the scientific basis of agreement. The International Energy Agency's World Energy Outlook published late in 2011 stated that if current trends continued we should have used up what then remained of our available carbon budget for keeping within the margin of climate safety over the next few years. Most recently (at the time of writing) the IPCC estimates that to have a better than evens chance of keeping to less than 2 °C of warming, further anthropogenic emissions will need to stay within an upper limit of some 1,210 billion tonnes of CO_2 equivalent. The current annual emission rate is about 50 billion tonnes according to Lord Stern, former World Bank chief economist and author of a landmark report on the economics of global warming.[4] At that rate, therefore, we have some 25 years' worth of carbon budget left. That is, we have just over two decades to decouple the global economy from carbon emissions altogether if it is to keep functioning at its present levels of activity.

The plain fact that this window looks now effectively closed underlies Premises 3 and 4. Given that despite repeated pledges by governments across the world, greenhouse gas output is still *rising*,[5] while successive international conferences have failed to get anywhere near the required binding agreements for starting to reduce it, turning annual emissions of 50 billion tonnes around in little more than two decades is in practice simply unfeasible. (And this would be so even without the explicit refusal of China, reported at the same time as Lord Stern's comments, to contemplate reducing its reliance on fossil fuels within the necessary time frame.) We can thus appreciate why the respected policy analyst Clive Hamilton, Professor of Public Ethics at the Australian National University, in his book *Requiem for a Species*, published in 2010, commented that:

> Until recently … anyone predicting the end of modern civilisation was arguably guilty of exaggerating the known risks, because the prevailing warming projections indicated there was a good chance that early action could prevent dangerous climate change. But in the last few years scientists' predictions about climate change have become much more certain and much more alarming, with bigger and irreversible changes now expected sooner … catastrophic climate change is now virtually certain.[6]

And for an example of one of those scientists, we could note Kevin Anderson of Manchester's Tyndall Centre, telling us with chilling sobriety in research published by the Royal Society also in 2010, that 'there is now little to no chance of maintaining the rise in global surface temperatures at below 2 °C … moreover, impacts associated with 2 °C have been revised upwards so that 2 °C now represents the threshold of extremely dangerous climate change'.[7] It is worth dwelling just for a moment on the uncompromising clarity of that informed scientific assessment: a rise of over 2 °C triggers extremely dangerous climate change, and we are now virtually certain to have a rise of over 2 °C. Not much wiggle-room there?

And yet this is, as I say, a crisis of recognition – because, confronted by failure of this order, very many serious and environmentally concerned people have gone on refusing to admit that they see it. A suitably Orwellian slogan – 'Failure is Success', perhaps – cries out for inscription, not just on the façades of environment and climate change ministries worldwide, but also on the letterheads of all those NGOs and activist groups who persist worthily in campaigning as if this crucial window for effective action had not closed. And yet all the while, the inadmissible awareness that this is, in truth, what has happened and what we are doing in response, has been growing more and more painfully insistent.

Copenhagen 2009: moment of truth? – or not

Denial, in other words, isn't just something the bad guys do. We are familiar enough with the idea of denial in the environmental context, but we have to recognise that it is to be met with not only at the soft hands of corporate power or in the soft heads of the US Bible belt. Nor is it a matter just of the ordinary, everyday shifting of the uncomfortable towards the back of the mind which people of goodwill, aware of the problem, nevertheless find necessary in order to get on with their lives. Denial is very firmly embedded in the thought and practice of those trying hard to save us.

Full recognition of this, for me, came with the UN's Copenhagen summit held in December 2009. Trying to get a bearing on denial, it is worth glancing back over how this epochal meeting of world leaders was seen beforehand, and how environmentally committed people have reacted to its aftermath.

That the almost millenarian anticipation that preceded that summit is now difficult to recall, is actually a significant datum. Obliged by an office move somewhat later (and just when I was beginning to think about writing along these lines) to sort through all those interesting documents which accumulate in corners, I chanced on a copy of *The Guardian* for 8 December 2008. 'Planet under pressure', reads the relevant headline: 'This week, ministers and officials gather in Poznan at the start of a one-year countdown to the Copenhagen summit, *at which experts say a deal must be reached if we are to have a chance of averting catastrophic warming*' (my italics). The key issues, further analysis notes, are signing up to a global goal for maximum temperature rise (or atmospheric carbon equivalent), and agreeing which nations will cut their carbon emissions by how much and by when. As I have already remarked, it is clearly stated that 'to be meaningful, targets must be short term, perhaps something like 25–40 per cent by 2020 for the developed world'.

Well, as we know, *that* didn't quite happen. My concern now, however, is not with the gross derelictions of global statesmanship on display in Copenhagen or since. It is rather to flag up that back in 2008, if we are accidentally prompted to recall it, *Copenhagen was the last-chance saloon*. So it is sobering to compare the terms of this cutting with the kind of language being held on the same topic a good deal more recently, well after Copenhagen had concluded in angry muddle and dismay. Here for instance is an address by Christiana Figueres, Executive Secretary of the UN Framework Convention on Climate Change, to a conference of Spanish defence and security luminaries in February 2011 (a full fifteen months after the Copenhagen fiasco). Putting the best face she can on the intervening follow-up meeting at Cancun, she still has to admit that:

> more is needed ... the level of ambition currently on the table amounts to only 60 per cent of what is needed to limit the temperature increase to the agreed 2°Celsius ... similarly, no agreement was reached on the year in which global emissions need to peak.

She continues dramatically:

> Ladies and gentlemen, this should be a wake-up call to the world! For
> if we do not manage to constrain carbon to the recommended level,
> we will collectively lose the ability to turn the mitigation key, and miss
> the opportunity to put the world on a pathway that does not hold a
> large potential for conflict.[8]

Or again, two years further on: when atmospheric CO_2 concentration was
reported in 2013 to be on the point of passing the symbolic level of 400
ppm, Tim Lueker, an oceanographer and carbon cycle researcher with
Scripps CO_2 Group, was quoted thus: 'The 400 ppm threshold is a sobering milestone, and should serve as a wake-up call for all of us to support
clean energy technology and reduce emissions of greenhouse gases before
it's too late.'[9]

And when the IPCC's Fifth Framework Report was published later in the
same year, the secretary general of the World Meteorological Organisation,
Michel Jarraud, called it ... yes, you guessed: 'another wake-up call that our
activities today will have a profound impact on society, not only for us, but
for many generations to come'.[10]

The point needs no labouring. If each successive missed deadline,
breached threshold or authoritative warning is a wake-up call to the world,
the global alarm clock must be set very firmly to 'Snooze'. And by the same
token, if the last-chance saloon turns out in practice to be the last-chance
saloon-car, always capable of being shunted on a few miles further down
the track without ever encountering any buffers, then its effectiveness as a
venue for concentrating the mind is hugely diminished.

What we find, here and very widely across the board – these comments are, as anyone can easily confirm, wholly representative – is not
insight and acknowledgement, but something else: a form of refusal to see.
And since Copenhagen, the UN especially has taken this refusal to levels
that would be comic if the issues weren't so serious. In preparation for
the Rio+20 Conference in the summer of 2012, the UN Environment
Programme produced a statistical snapshot showing that in the twenty years
since the 1992 Earth Summit, CO_2 emissions had increased by 36 per cent
and atmospheric CO_2 concentration by 9 per cent , reflecting among other
things a world increase in energy and heat generation of 66 per cent, a
doubling of passenger trips by plane and a global population growth of
1.5 billion. Over the same period, it shows the global average temperature as having increased by 0.4 per cent and the ten warmest years since
records began as having all occurred since 1998.[11] And in response, amidst

the portentous smother of words constituting the Conference's concluding resolution: 'We recognise that the twenty years since the United Nations Conference on Environment and Development in 1992 have seen uneven progress'[12] – a remark surely ranking with Emperor Hirohito's observation after Hiroshima and Nagasaki that the war wasn't really going Japan's way. Yet later, if one can bear to read on, one still finds the now-routine battle cries of commitment and determination: 'We reaffirm that climate change is one of the greatest challenges of our time, and … we emphasise that adaptation to climate change represents an immediate and urgent global priority'.

A willed refusal to confront reality powers this approach. Admitting that destructive climate change is coming is taken to mean despair, which would paralyse us. But refusal of this paralysis, we are overwhelmingly encouraged by contemporary civilisation to believe, requires *optimism*. In the climate change context, that means soldiering on from defeat to defeat, believing resolutely that we can still save the world for continuing human betterment. Therefore, if we want to act at all – and how could we just give up? – it *can't* be too late for that. Therefore, it isn't.

What if we looked down?

All this could be conceded, and I could still devote a whole book to probing Premise 3 of the Vicious Syllogism, which is where wiggle-room for avoiding the conclusion would have to be found if there were to be any at all. Is the window now actually closed, such a book might enquire, or just closing with frightening rapidity? A self-imposed global regime of the comprehensiveness and severity required may be utterly unprecedented in the whole of human history, and from an economist's or social psychologist's perspective barely imaginable, but not even that makes it literally *impossible*. So maybe, just maybe … And if I were to take this line, I should be in numerous company, some of it distinguished – such as that of Chris Rapley, Professor of Climate Science at University College London, whose Foreword to a recent book claims that 'on an otherwise finite planet, our ingenuity, at least, seems unbounded. We still have the chance to limit the scale of the disruption.'[13]

I shall not, however, do this. Overcoming huge reluctance, after half a lifetime's environmental activism in various modes, I find myself in the position represented by my epigraph from Nietzsche. I have come to believe, and to recognise that I have actually believed for some while, that the conclusion of the Vicious Syllogism is true – we are, already and inescapably, in for a very rough ride indeed from oncoming climate change.

You may dispute that belief, or recoil from agreeing with it. But whatever your reaction, I am going to suggest that for the purposes of this book you just grant me it – for the sake of argument, if you like. Because it seems to me that very many of the environmentally concerned community are simply refusing to confront it as even a real possibility, and I also think that a lot of this refusal stems from a kind of conceptual vertigo – we keep on pretending because we daren't look down, for lack of any way of thinking how we might stop pretending and still keep our footing. So I want to ask, as the poet Paul Kingsnorth, chief instigator of the Dark Mountain project, asks in its powerfully written manifesto: '*What would happen if we looked down?*'[14]

Suppose we were indeed to stop pretending. That would mean facing up in full knowledge to an ugly and very dangerous future, both as this generation's legacy to its successors (a profoundly humiliating thought, or it should be), and as what will increasingly become the background of our own and our children's lives. It would also mean admitting something that is, if possible, even more difficult to confront: that it is *our* lives, yours and mine, which are bringing this havoc ineluctably on. We – those of us of riper age, especially – should have to live with being the generation that has jeopardised life on Earth.

But our lives would meanwhile have to go on, and for that to be possible, despair couldn't be a final option. As Kingsnorth well says, anyone who doesn't feel despair in times like these is not fully alive[15] – but that couldn't be the last word. Being fully alive also means not simply sinking into resignation or apathy. So what *could* we do? And prior to that – importantly prior to it, and not just for someone like me writing as a philosopher – how do we *think* about what we would then be confronting? For how we think is not something separate from how we act: we can only act through the concepts that frame our understanding, because only in terms of how it is conceptualised can anything *be* a purposive action.

A very powerful conceptual framing through which we currently act, by taking it unquestioningly for granted, is the idea that for life to go acceptably on, we have to be pursuing some kind of *progress*. This ruling idea ensures that we expect attention to the human future to fit the rubric supplied by the American historian of progress Robert Nisbet: 'mankind has advanced in the past … is now advancing, and will continue to advance through the foreseeable future'.[16] Advancement in this context means improvement in the general conditions of life driven by the cumulative growth of both material wealth and scientific knowledge.

The power of that expectation is clearly on display, for instance, in a recent book by the doyen of British environmental campaigners, Jonathon Porritt. *The World We Made* is a sort of 'Age of Smart' to set against Franny Armstrong's 2009 film *The Age of Stupid*. It looks back from the future to tell the story of how we got our world back from the brink of collapse to 'where we are now in 2050'. Key game-changers that Porritt thus 'retrospectively' envisages are an energy security impulse in the US carrying to office a Republican president who drives through a carbon emissions permit scheme; an Emergency Report by the IPCC (about which nothing is said except that the findings were 'devastating'); an internet-enabled worldwide youth rebellion in response to a World Bank report on widening wealth disparities, rocketing youth unemployment and the IPCC Emergency Report, leading after several months to government action across the world on these issues; and a decision by the Roman Catholic Church to make birth control an issue of individual conscience, leading to a global population downturn. These changes of collective mind and direction are played into a context of continuing hi-tech developments driven by an increasingly environmentally responsible capitalism, achieving *inter alia* 90 per cent of world energy production from renewables by 2050, and very significant carbon capture and storage worldwide. It is, as he frankly admits, a 'technotopia' – a benign *Brave New World* including empathy drugs, artificial meat, wired-up digitalised health monitoring and complete internet connectivity facilitated by brain implants.[17]

It would be flat-footed to get into an argument about the specific plausibility of any of this. Porritt is as impressively well-informed and forceful as ever, the book is full of interest, and optimism always has all the best tunes (or at any rate, the catchiest ones). What we need to note in the present context is the aspiration that concludes both the Introduction and the final chapter, and clearly frames the whole thing:

> we have confidence, once again, in our shared ability to make a better world for all those who come after us.
>
> We're not there yet … but we're making good progress.[18]

The note of his peroration, just prior to that last claim, is in fact this: 'If it's a better world we're after, just make sure that every child reaches the age of six feeling radiantly happy.' As I say, all the catchiest tunes … but here, surely, the catchiness has a Disney feel to it, so blithely sentimentalising is that view of what life is about. The point can't really be universally happy kids, nice though that would be – it is much more like: struggling, as grown-ups, to live from the whole of ourselves and to help our children do likewise. Yet, as a criterion for positive action, happy kids is really no more sentimental than 'a better world for all those who come after us'. In either case, warm glow is made to carry much more weight than it deserves by being projected forward in the guise of altruism. As the generations march onwards and upwards, each bettering the world for the next, one is supposed to feel that life is thereby justified. But again, no real *purpose* can be manufactured out of bettering the opportunities for the next generation to better the opportunities for … and so on. For all its up-to-the-minute air and its lively appeal, *The World We Made* is actually a classic of progress as 'social hope'. To say this is in no way to belittle a fascinating book, still less its author. On the contrary: it is to demonstrate the hold that the 'progress' mantra can obtain even over the most distinguished environmentalist of his generation.

But if the Vicious Syllogism is right – which is, to put it no higher, at least as likely as Porritt's game-changing scenario – progress is over. Expectations of a continually bettered world are not going to hold any longer, and the catchiest tunes are being whistled in the dark. Through what conceptual adjustments, therefore, could we prepare ourselves to adopt something that might count as a positive attitude towards the ending of such expectations? I don't say to welcome it, which would be just too grotesque, given how drastic a future this is likely to be; and I certainly don't mean the sort of glib and facile speculation about new opportunities for growing oranges in Northumberland and so forth, that one sometimes sees. But might there be a posture of openness towards our responsibility both for climate change and for what emerges from it, in which we could at least *befriend*, as it were, the inevitable?

Very importantly, that doesn't mean mere passive acquiescence. Looking down is not giving up, and the end of the world as we have known it need not be the end of the world. What is at stake is the spirit in which we do the various things that may still be worth doing, both to mitigate the effects of climate change and environmental damage where we can, and to retrieve what is possible of humane living from the forces that are bringing these about. If we go on in the spirit of denial, the spirit of regretting progress as perpetual advance and trying to retain as much of it as we can, we shall not be learning in a sufficiently chastened way from our mistakes. We need, as Freud might have put it, to be able to let go of this attitude in order to mourn it (perhaps), and then move on.[19] But progressivism – unquestioning attachment to the idea and experience of progress – has a very powerful grip on our minds. So what could letting go involve?

My approach to answering that question has four guiding themes:

Tragedy, not problem: We must stop seeing destructive climate change as something accidental to the human condition, something to twist and contrive and technologise our way out of in the nick of time if we can. Instead we must come to recognise it as something that deeply flawed human strengths have brought on us as an almost inevitable corollary of their deployment, something that we must simply try to come through and from which the best we can do is *learn*.

Now, not the future: In large measure to avoid that difficult recognition, we have taught ourselves to think of environmental damage as something we are mostly doing to our successors, to future generations – although the first foreshadowings of these effects are already with us, in what is happening to the Arctic hunting grounds of the Inuit, for instance,[20] or to low-lying island nations such as the Maldives. But we also need to understand it as, in the first place, something grievous that we are already doing to all of ourselves, here and now in the comfortable West and in the present. This means freeing ourselves from the distracting late-twentieth-century mindset of 'sustainable development', with its obsessive focus on inherently negotiable futures, and turning again directly to the issue of how we are currently travestying our ongoing living relations with the natural world around us.

Wildness, not well-being: Insofar as sustainable development thinking sees those relations in any other terms than resource usage impacting adversely on the future, it does so with the claim that rectifying them promotes present human well-being – the so-called 'quality of life' agenda, according to which (for instance) insisting on organic carrots from the supermarket promotes the health both of oneself and of the planet. But this begs a lot of hard questions about life quality: and even if we could find agreed answers, it would be much less important than the issue of recovering human *wildness* – that is, our ability to live and act as whole natural beings, undisintegrated and unalienated.

Hope, not optimism: Denial of our real situation can be directly self-serving, as when commercially vested interests are at stake, but it can also express and seek to protect a forced optimism that we take to be the only alternative to despair. But the opposition that matters is that between willed optimism and *hope* – which, being necessarily unwilled, is at bottom the essence of human wildness. Only hope, not needing to base itself on pretending, can accept the tragic – and thus have any chance of seeing us through.

That is meant to give the flavour of the book. For those with strong enough minds and stomachs still to read on, I will indicate briefly how the following chapters interweave and develop these themes.

Tragedy, not problem

The difficulty here is that while problems can sometimes have solutions, tragedy involves terrible loss, unmitigated and uncompensated.

Terrible loss: let's just get that clear. Even if the worst-case scenario of runaway warming up to 6 °C and beyond does not materialise, the increasing climate instability to which we have already committed ourselves will certainly be accompanied by rising sea levels, extensive desertification and ecosystem breakdown worldwide. While we cannot know with any precision what that will mean in geopolitical terms, some general features are horribly predictable: there will in many parts of the world be famine, epidemics and homelessness on an epic scale. We face by 2030, a former UK Government Chief Scientific Adviser has warned, a 'perfect storm' of food, water and energy shortages.[21] Consequent on these developments will be enormous migrations, or attempts at migration. Those currently temperate parts of the world where the immediate climatic effects of warming are likely to be comparatively less drastic will come under enormous pressure to admit refugees in numbers that would quickly overwhelm their resources and infrastructure. They will need, both for their own survival and as the only trustees of civilisation left, at some point to close their borders. The consequence of this, in turn, will be inter-communal and international conflict, and much of this conflict will equally inevitably be armed. The world is set to become not just a significantly less habitable but a vastly more divided, hostile and violent place. Gwynne Dyer's book *Climate Wars* explores a range of possibilities here, most of them plausible and all of them deeply disturbing. If *any* of them is realised, the human project is going to judder grindingly into reverse.[22]

What we have mostly done with this prospect so far is refuse to contemplate it. In the process, we have illustrated how easy is the transition from refusing to accept that there are problems, to refusing to accept there are anything *more* than problems. In Part I of the book I probe this reaction of denial – one that is, for all except the stolidly ignorant, essentially in bad faith. Chapter 1 identifies the different forms that this can take, and looks at examples of each (including, as indicated in the Preface, my own earlier work). It accounts also for key features of the sustainable development paradigm in these terms. It emphasises, as something needing urgently to be explained, the fact already noted, that precisely the same patterns of denial are to be found among those who fully accept the reality of climate change and ecological disruption as among those who want to shy away from it.

Clive Hamilton calls his *Requiem for a Species* 'the story of a battle within us between the forces that should have caused us to protect the Earth … and those that in the end have won out'.[23] The history of that battle is largely the history of the idea of progress, which I trace in Chapter 2, showing how sustainable development seeks to continue material progress by capturing it for 'ecological modernisation'. That perverse endeavour is, in turn, evidence of how deep the hold that progress has on us actually goes. In Chapter 3, I try to track the roots of this *progressivism* down to our inability to confront our own mortality and finitude – to recognise ourselves fully, that is, as organisms in nature, wholly organic forms of conscious life.

It is only at the level of what might be called existential failure of this kind that we begin to approach tragedy proper, the kind of inner battle where disaster ensues from and expresses destructive weaknesses that are *structurally inherent in key life-strengths*. Understanding this pattern requires an account to the structure of human being, to which I turn in Part II of the book.

This understanding offers our only grounds for hope. Refusal to recognise environmental tragedy rules out any possibility of real learning. But if we actually face up to the fact that our situation is tragic, and ask ourselves what has gone so *deeply* wrong with the character of our present way of living as to have ensured oncoming climate destabilisation and gross ecological damage, we might be able to use the answers in retrieving from it what we still can. Part III is occupied with providing some thinking tools for exploration here.

Now, not the future

We can't understand what is genuinely tragic about our attachment to progress – how good and evil have been genuinely interwoven in it – without bringing in the dimension of time, and specifically of present vis-à-vis future evils. Considering sustainable development thinking as a paradigm of denial, as I do in Chapters 1 and 2, already has a clear relevance to this theme. For sustainable development deflects attention from what is really wrong with us in the present by concentrating on the future consequences of our actions, and on the obligations of justice to future people that in preparing those consequences we are allegedly breaking.

But we need to revisit the history of environmental concern in order to recover a sense of what this paradigm has perverted. This I do in Chapter 4. As I indicate there, the concern that was originally strong in both the

British and the US environmental traditions was not, on either side of the Atlantic, about 'the future' in the ecological sense, but about what people then alive, and their children growing up, needed for a fully human life. It was premised, in fact, on recognition that, as John Muir put it, 'wildness is a necessity'.[24]

Correspondingly, the central responsibility that classic environmentalism accused modern civilisation of neglecting and betraying was not to future people, but to themselves – to those alive in any given present, whose need for wildness is a need to realise the full organic-conscious unity of their species being. Chapter 6 offers an account of the existential grounds of this crucial natural responsibility.

Wildness, not well-being

On this account, engagement with wildness as manifested in the outer world – with what lives and acts only from the whole of itself, as natural force – is a necessary form of deep intelligence, without which human beings can't make sense of themselves (in particular, of their mortality) and are thereby tragically locked into a futile, escapist and destructive progressivism. It is a necessary form, as I try to show in the central chapters, because the deep structure of reflexive consciousness, the conditions of our unique epistemic power, occlude and screen from us our natural unity as finite organic beings. But we can tacitly re-inhabit that unity through re-weaving the symbolic (and much more than symbolic) wildness of the outer world back into our patterns of life.

We need wildness, that is, not to make our lives go better, but to enable them to *be human lives* – as the ground for all better or worse. So this is not about how we can ensure ourselves maximally rich and satisfying life-experiences by turning our aspirations from material consumption to the pursuit of genuinely satisfying life quality decoupled from ecologically destructive pressure on resources – important though doing that undoubtedly is. Without living from our wildness, we can't be reconciled to our finitude, and so we can't make sense of being alive, and are thereby committed to a perpetual treadmill of impossible escape from that senselessness – the perpetual pursuit, from one project of self-betterment or of social progress to the next, of *ersatz* meaningfulness which we know all the time to be *ersatz*, and so can only pursue all the more desperately. (This applies just as much, be it noted, to the quality-of-life agenda.) And living from our wildness means direct engagement with whatever forces us to recognise that we are not only not in control, and could never really be in control, but that we actually flourish in the life-acknowledgement of our not being in control.

With all this, too, I am trying to grapple in Chapters 5 and 6 of Part II, and also in Chapter 7.

It is these central chapters that are most likely to try the patience of the reader without a philosophical background. But the work that I have tried to do there is essential to the argument. All I can plead is that I have tried hard, as in *The Sustainability Mirage*, to do this work directly and at first hand, assuming no prior philosophical familiarity.

Hope, not optimism

If we can reawaken natural responsibility as a perceived human need, retrieving it in practice could actually be prompted by the coming disintegration of the global-megalopolitan machine, providing that we can carry enough of viable civilisation through. And this is a genuine, though inevitably a rather bleakly exposed, ground for hope. In Part III of the book I consider what might be built on that ground.

How might we re-engage with wildness, genuinely and without sentimentality? This is the burden of Chapter 7. We call climate change *anthropogenic*, and then in our arrogance take that to mean man-*made*. But it is not man-made, any more than the tree that you plant as an acorn and which then leaps up to its own self-shaped maturity is man-made. Oncoming climate change is man-, or human-, *occasioned*, human-induced, human-*triggered*, at most. And what is thus triggered, the forces thus released, come to us out of the wild world both beyond us and including us, as surely as have a million springs. So one hope of retrieval may be to recognise and (if we can find it in us) to accept positively, this irruption of the ineliminably wild back into lives that had forgotten it.

But hope is *not giving up on life* – and confronting realistically the prospect that it is now almost certainly too late to save the Amazonian rainforest, or 50 per cent of European biodiversity, or the Maldives, or sub-Saharan Africa, looks like giving up on huge swathes of life. What could hope find to say in face of anything involving such an affront to what E.O. Wilson calls our instinctive *biophilia*,[25] and in human terms so horrendous?

This question it is now above all the role of the environmentally-intelligent community to answer. The issues here are challenging. How could we continue to reduce environmental impacts as part of a survival drive towards resilience rather than 'sustainability'? How could an important place for high technology be retained? (What about nuclear power, for instance?) What about education, food and energy production, transport and other infrastructures? In politics, are environmentalists really just a cuddlier version of liberal democracy, or a vanguard party? If the latter, what would the role of a green vanguard be in organising the construction of a viable alternative local infrastructure, using the means and people available? What alliances (on past form, perhaps, quite uncomfortable ones) will there need to be on the national scene, in particular over maintaining the essential capabilities – including the armed security – of the survival state? And how can we justify doing only what we can where we can when so much beyond our borders will be in chaos? Chapter 8 offers not easy answers, nor even difficult answers, to such questions, but some conceptual equipment for starting to explore the most fundamental of them in the spirit of existential resilience.

It is clear even from an initial exploration that as well as some disconcerting new departures, there is likely to be a good deal of continuity here with what many are already doing. Even so, a changed framework for understanding what we must carry on doing changes the lights and resets the possibilities.

But this is not optimism. Optimism, if you once admit the Vicious Syllogism, can now only mislead. Once you have accepted that a massive, decisive and in effect immediate shift to a sustainable society based on renewables and reduced consumption – the old Green dream, even when given a shiny new technological gloss – just isn't going to happen, then to remain an optimist you will have to see humans assuming some sort of technologised God-role. In this perspective, because failure to 'save the planet' is inconceivable we must be in time, even at well past the eleventh hour, to succeed – through as many massive and implausible geoengineering projects as are necessary. As Mark Lynas says at the close of his most recent book, *The God Species*: 'Only optimism can give us the motivation and passion we will need … Voices of doom may be persuasive, but theirs is a counsel of despair. The world – and our children – deserve better.'[26]

But that opposition between despair and optimism is false. Willed optimism, kept afloat by denial, is not an alternative to despair but a form of it. All turning away from the real is despair. Of course, recreational fantasising can be harmless enough: but fantasy in place of the real is giving up on what, after all, is all there is. Correspondingly, the opposite of despair is not optimism, which so readily betrays us into bad faith, but hope – which must always, again by its nature, be an embracing of reality.

The question whether humanity can indeed learn such hope from confronting its current plight is explored in the brief Coda that concludes the book.

Beyond pretending

What I offer here is avowedly not a blueprint. (One of the things we will have to stop pretending to have, outside engineering, is blueprints.) Rather, it is a way of thinking ourselves through to new understandings and options. Only when *What could still happen after the end of pretending?* is a question genuinely open, can we find it in us to respond with something like this: We know a smash of some sort is coming. We don't know how we will emerge on the far side of it, or even whether there will *be* a far side. But by the same token, we don't know that there won't be – at least for some communities in some parts of the world. This is a tragedy that humanity must face with grim regret for what it has brought on itself; but whether civilisation in any recognisable form does come through it will depend in large measure on the resources that we can now find within ourselves to take into it. What can we do now, in our living and working arrangements, in our necessarily continuing cohabitation with a biosphere that we have so extensively despoiled, to retrieve and cultivate what we can?

If our approach remains configured as optimism, by contrast, or even as meliorism – least-worst-ism, the aspiration to mitigate and adapt so that *something* recognisable as 'progress' can still be pursued – then we shall remain wide open to all the solicitations of denial. And those solicitations are insidious, potent and treacherous. Sustainability, like the mirage it is, has failure to reach it built into the pursuit of it. That, for corporate power and the mainstream political charade, is its point – that's why they invented it. But if those who stand up on behalf of humanity allow the same way of thinking to warp their life-hope as well, that endless pursuit will simply waste our remaining time for action.

This book is therefore concerned with the conceptual work needed for building strength to look at our situation straight. I think of that strength, for reasons that will appear, as *existential resilience* – a kind of robustness required for being fully ourselves. We in the West and North have lost that, as we have lost more obvious kinds of robustness, in the process of developing over the last three centuries our urbanised way of life, with all its increasingly smooth technical facility, and imposing it more and more widely across the world. There is, as people always insist when this kind of point is made, no going back. But once we have understood what is involved, there are possible ways of going creatively forward, even through breakdown and grave damage. And a recovered existential resilience is needed to inform the various on-the-ground forms of resilience that we will have to rebuild if we are to retrieve anything worthwhile from what is coming. To repeat, how we think cannot be separated from how we act: the conceptual and the practical are not fundamentally distinct.

The future will certainly be grievous. We can safely guess, however, that soaring temperatures, drought, crop failure, famine, desertification, the incineration of forests and rising sea-levels, will not strike with equal effect everywhere all at once. In some of the places where they will not strike, or strike catastrophically, until well into the medium term, there remains the cultural and institutional potential to break free, even at this terribly late stage, from the 'progress' mind-trap, and build a new understanding of *humanitas* for whatever *afterwards* there may be. Nor is it foreclosed that there could be an afterwards, though we must stop trying to foresee it in any detail. We don't know, for instance, and shouldn't expect to be able to know, what implosion of the global capitalist economy might mean by way of capping emissions far more effectively than any UN agreement. We don't know how soon, or in what ways, that process may feed through to lessen the risk of catastrophic runaway warming, if indeed it does. Given where we are coming from, probably the hardest thing for us to accept is that it is not our business to know these things. It is our business to learn as best we may from the mess we have made, and to keep the human torch alight meanwhile, if we can.

That is at any rate the work on which this book is engaged.

Notes

1 Nietzsche (1906/1968), sect. 25.
2 Lynas (2007), p. 270.
3 For targets at Copenhagen, see 'Planet under pressure', *The Guardian*, 8 December 2008. For the IEA World Energy Outlook 2011, see www.iea.org/publications/freepublications/publication/WEO2011_WEB.pdf. For the IPCC Fifth Assessment Report, see www.ipcc.ch/report/ar5/wg1/
4 As reported in *The Guardian*, 30 September 2013, 'Top economist issues 25-year climate warning'.
5 As documented for instance in the IPCC report just cited.
6 Hamilton (2010), p. xiv.
7 Quoted in *The Guardian*, 29 November 2010, 'Climate change scientists warn of 4C global temperature rise', www.guardian.co.uk/environment/2010/nov/29/climate-change-scientists-4c-temperature
8 Address available at http://unfccc.int/files/press/statements/application/pdf/speech_seguridad_20110215.pdf
9 Quoted in *The Guardian*, 30 April 2013, 'Global carbon dioxide levels likely to hit record levels' (sic).
10 As reported at www.bbc.co.uk/news/science-environment-24295907
11 UN Environment Programme report *Keeping track of our changing environment* accessible at www.unep.org/GEO/pdfs/Keeping_track.pdf

12 *Resolution adopted by the UN General Assembly 66/288: The future we want* (paragraph 190), accessible at www.uncsd2012.org/thefuturewewant.html
13 See his Foreword to Weintrobe (2013), p. xx.
14 For the Dark Mountain 2009 manifesto *Uncivilisation*, see http://dark-mountain.net/about/manifesto/
15 See his piece 'Dark Ecology', available at www.paulkingsnorth.net/journalism/dark-ecology/
16 Nisbet (1980), pp. 4–5.
17 Porritt (2013).
18 The quotations are from the Introduction (p. 6) and the final chapter, 'The Great Turning' (p. 270).
19 See the well-known paper 'Mourning and Melancholia' – Freud (1961), Volume XIV, pp. 237–58.
20 As well discussed, for instance, by John Broome in Broome (2012), ch. 1.
21 See the report in *The Guardian* for 18 March 2009 ('World faces "perfect storm" of problems by 2030, chief scientist to warn') of a speech by Professor Sir John Beddington, UK Government Chief Scientific Adviser 2008–13.
22 See Dyer (2008).
23 Hamilton (2010), p. xii.
24 Muir (1901/1997), p. 721.
25 See Wilson (1984).
26 Lynas (2011), p. 244.

PART I
Denial

1

VARIETIES OF DENIAL

Denial is, as we have seen, an environmentally crucial issue. But we need to be clear what is being meant by the term.

There are of course still a few people who will deny that humans are decimating the biosphere and warming up the atmosphere, simply because they believe these things not to be the case. That, however, is a position which by now, in a wired and mass-communicating world, can only really be taken by the ignorant or the misinformed. The extent and direction of cumulative adverse change have now been so plainly evident to the relevantly expert eye for so long, while the scientific, economic and socio-logical explanations that decisively relate this to human activity have also been so widely available, that deniers in this very crude and basic sense no longer have any vestige of an excuse.

Human nature being what it is, the ignorant will always be around. But it is not with them that the real trouble lies. After all, in the ordinary way of things, ignorance is removed by correct information combining with appropriate experience. If, however, simply telling people the truth about what they can increasingly see to be happening to the world were all that was required, we might have expected there to be a far more active accept-ance of the need to make radical changes to Western lifestyles in the face of global warming, in particular, than any evidence suggests there has been.

The forms of denial that should worry us are those that are in one way or another *willed* – those where what is denied is also, and at some other level, recognised to be true, or to have a good chance of being true: where denial is not straightforward, albeit misplaced, belief to the contrary, but explicit rejection of what is tacitly admitted but unacceptable. Such denial is not amenable to being overcome by improved access to the facts – exposure to the facts can, indeed, tend to strengthen it. That is because willed denial always goes with vested interests of one kind or another. It is always a form of refusing to know something that really we do know, or have at any rate very strong grounds for suspecting, because we have too much staked on not admitting it. It has been recognised for some time that this reaction is peculiarly prevalent among those who are *not* ignorant or misled about anthropogenic climate change.

Denial: literal, interpretive and implicative

As such, willed denial takes various forms and operates at different levels. These can be explicated effectively within a framework originated by the late Stanley Cohen.[1] A sociologist, he develops his analysis in the context of reactions to atrocities, human rights violations and other kinds of abuse, but it transfers readily to environmental situations, and in particular to that of climate change.

Cohen distinguishes *literal* from what he calls *interpretive* and *implicatory* denial. We could sketch a quick preliminary distinction among these varieties as follows. Literal denial says, of something tacitly recognised as a cause for concern, that there is nothing really the matter – appearances here are deceptive and actually, on proper scrutiny, there's no problem. Interpretive denial says, usually more plausibly, that while something is indeed the matter, it's not actually that serious – what is at issue has been wrongly interpreted, misconstrued, maybe blown out of proportion by doomsayers, but there's nothing here we can't deal with. Implicatory denial pursues the same goal of reassurance, but by a different route. It says in effect: OK, it's serious – let's not go there. This recourse clearly cannot be adopted as blatantly as that. Indeed, for it to work at all the fact that it is being resorted to must not be crudely apparent. But changing the subject, that is, quasi-intentionally not following up on the uncomfortable implications for thought and action of the admission that things are serious, is nevertheless the essential tactic.

How it is that the human mind can play this kind of trick – can simultaneously both recognise and avoid recognising the same thing to be the case, while also both knowing and not knowing that this is what it is doing – is a fascinating study in itself. Attempts at explanation have included the existentialist philosopher Jean-Paul Sartre's classic account of bad faith – 'a certain art of forming contradictory concepts which unite in themselves both an idea and the negation of that idea' – as a fundamental potentiality of human consciousness,[2] and also the Freudian concept of disavowal as a mode of psychic defence against traumatic experience and anxiety-inducing perceptions.[3] I have myself invoked the former, and others more recently the latter kind of account, in relation to our stance towards climate change.[4] We shall need to return to this briefly at the end of the chapter. Meanwhile, we must see how these three forms of denial are actually represented in the climate change arena.

About literal denial there is no call to say much here. This appears in our field as the assertion that anthropogenic climate change isn't really happening. It is different from mere substantive ignorance, and much more reprehensible, because it involves identifiable processes of active *ignoring*. It is most clearly evident when a great deal of well-funded ideological effort (in the service of very obvious corporate vested interests) goes into rubbishing the relevant science and encouraging climate change scepticism. Literal denial here involves the choice, which will always at some level be a matter of willed inclination, to get one's 'information' on the topic from this kind of source, or from the hired press which lackeys it. We already know quite enough about the bad guys. But much more to the point, as I suggested in the Prologue, is to explore how denial has also been apparent on the side of the angels.

Not that some of the bad guys haven't visited the reformatory, albeit under compulsion. Closely related to literal denial here, and often driving the disinformation that facilitates it, is refusal of responsibility – a self-interested turning away from the climate issues by those hoping to profit from carbon-intensive activities in the short term and leave the mess for successor generations to deal with. One very notable achievement of the environmental movement to date has been to make this irresponsibility harder to get away with than it used to be. Even corporations can be shamed, and a good many whose strategy this was (the naming of names would for obvious reasons be ill-advised, but readers will be able to supply their own instances) have now been forced, either by consumer pressure, shareholder activism or simply changing public attitudes, to move to a less exposed position.

That position, however, has typically been one of interpretive denial. They have joined the very numerous camp of those who admit the carbon threat, but deny that it represents anything beyond our technological and administrative capacities to overcome. In interpretive denial, the facts are not challenged but are interpreted in a way that makes them acceptable, or at any rate significantly less unacceptable: in Cohen's examples, 'I am a social drinker, not an alcoholic; what happened was not really "rape" ... this was population exchange, not ethnic cleansing; the arms deal was not illegal, and was not really an arms deal ... '.[5] As regards climate change, those now doing the damage or variously complicit in it, having been compelled to recognise that it is indeed damage that they are doing, clearly have vested interests in interpreting global warming as something essentially tractable and manageable. In particular, runs this interpretation, we can address and manage its consequences without having to change dramatically the structures and 'standard of living', so-called, of the advanced Western societies – to which we in those societies have so long been accustomed, and which the rest of the world strives increasingly to emulate. Those who profit from selling the blizzard of *stuff* constituting the standard of living, or the means to the frantic mechanical scurrying in the course of which we acquire and dispose of all this stuff, or the complex network of services that support the whole mad merry-go-round, have their vested interests in going on turning a quick buck, even if some of the bucks now have to be, or to be represented as being, 'green bucks'. And those charged with governing the societies and administering the economies have a vested interest in covering their backs by seeming to square up in a competent and hard-headed way to problems about which growing numbers of their more articulate constituencies are disquieted, while actually doing nothing that would make the whole business of daily modern life noticeably different.

Meanwhile, the majority of people, whose main vested interest is in getting through that ordinary business of life without being haunted by a huge background worry of this kind, rely complacently (when the issue is allowed to come up at all) on the reassuring thoughts that our scientists are marvellously inventive and nature is actually a lot more resilient than we suppose.

So far, so all-too-sadly familiar. As we shall see, however, that is not the whole story, and interpretive denial of the facts of climate change has lately acquired some unexpected new exponents.

'The social production of innocence'

Perhaps the most interesting place to start probing, however, is with an environmental example of Cohen's implicatory denial, where what are denied or minimised are not the facts or their ordinary significance, but rather the psychological, political or moral implications that would conventionally follow from such facts. A disturbing recent book has shown in close detail how a case-study community both widely alert to the issues and directly exposed to the early consequences of anthropogenic climate change, still deploys what are in effect collaborative forms of double-think to preserve the ways of life that it can see to be causing the damage.

In this study,[6] Kari Norgaard provides a thorough, careful and well-documented exploration of how various social processes implementing this order of denial operate. The case-study community is from Norway, which as well as being probably one of the best environmentally-educated societies in the world is (or was in 2009) the world's fifth largest exporter of oil and second largest of natural gas. It is a land, too, both of traditionally eager skiers and of winter snows which are – thanks to climate change – observably receding. Norgaard demonstrates how, faced with these tensions in theory and experience, members of a representative Norwegian community rely on a cultural toolkit of resources for negotiating them. She shows the community bringing to bear, through its ordinary daily interactions, social norms of conversation, emotion and attention to screen, by tacit common consent, what can acceptably be talked about, appropriately felt and brought into any shared practical focus.

Norgaard notes how 'ignoring something – especially ignoring a problem that is both important and disturbing – can actually take quite a bit of work'.[7] Active ignoring, we have already noted, is also a main feature of literal denial – but in this Norwegian case, the effects of climate change are simply too visible for their non-existence to be asserted, so the work involved is a lot subtler and requires a good deal of carefully inhibited self-awareness. In the nature of this work, it is not going to be achieved by the individual alone – it is bound to be a tacitly collaborative enterprise. With abundant illustration drawn from a considerable period of local engagement as a participant-researcher, the author shows this enterprise being carried on through observation of a variety of dialogical norms: the broaching of climate-change topics mainly at the level of small-talk, so that discussion cannot get too intense; confining political discussion to the local and do-able; ensuring that ways of raising the topic in educational contexts are positive and not disconcertingly scary for the kids; and keeping family and friendship conversational spaces undisturbed by 'external' worries. In parallel, norms of emotion emphasising the importance of control and self-possession (key traits, apparently, in the traditional Norwegian self-image) are deployed to mute the experienced intractability of climate change and to distance whatever cannot be brought within the domain of the rationally manageable. At the same time norms of attention – 'the social standard of "normal" things to think about' – construct a double reality in which climate change futures are acknowledged, but kept out of the sphere of 'real life' organised by predominant attention to the past and the present. All this is aided by an underlying mythic narrative of Norway as a small but well-ordered land of pristine nature and committed nature lovers, an image that plays a large part in maintaining what Norgaard calls, in a telling phrase, 'the social production of innocence' in relation to climate change.

But innocence here is a much more fine-grained condition than mere lack of knowledge or agency. What is socially produced under these pressures is not so much the suppression of well-attested information about causes and consequences, as the muting of its discordance with ordinary life-expectations in an advanced technological society; not so much the refusal to acknowledge collective responsibility, as the rendering acknowledged responsibility collectively *habitable*. The Norwegian locale and situation were chosen in large part for the peculiar starkness with which they bring out these features, but once alerted to them we can see that the issues are of quite general relevance. They transfer very readily, for instance, to the British context if we substitute slightly different norms and mythic references (the stiff upper lip, the amused, tolerant dismissal of eccentrics who take things too seriously, or the narrative of Britain as a liberal society with a long tradition of careful, adaptive assimilation of new ideas and developments). Social processes of implicatory denial, in fact, subsisting in culturally and historically varying patterns of relation with the literal and interpretive varieties, can be seen to characterise all those societies advanced enough for a strong leavening of climate change awareness to have permeated them, however that is expressed in their different political arrangements and intellectual profiles. Faced with a choice between, on the one hand, registering unequivocally what is both understood in principle and actually obtruding itself on their recognition, and, on the other hand, continuing in the courses of activity in which they have invested their lives, people find ways of both seeing and not seeing – and not just in Norway but across the Western and 'Northern' world, and increasingly elsewhere.

Nor is it just the essentially uncommitted who react thus: and here we come to a point of major importance. Paradoxically (at first blush), these ways of seeing-and-not-seeing often include engaging, and even engaging still more vigorously than before, in various forms of environmental activism at the level of small-scale local improvements in 'sustainability'. This is an obvious and on the face of it perfectly rational response to living in 'risk society'. As one of the people quoted in the Norgaard study puts it:

> It's like that with the environment ... there are lots of problems ... there is a lot that is negative; I feel a bit like, yeah, pffff! But when I have something that I am trying to do, ... when you are trying to influence something, then it's like you can be optimistic anyway.[8]

Wait a minute, though ... 'Think globally, act locally', goes the old environmentalist slogan. Doesn't the attitude expressed here evidence, not denial, but the robust pragmatism of getting stuck in to what you might be able to do something about at first-hand? Yes, but … that clearly grateful retreat to being 'optimistic anyway' should give us pause. Things are more complicated than they were in the 1970s. We must recognise that 'acting locally', when that means doing what I reasonably can in my particular situation (and who could ask more of me?), can now be a very good way to distance the acute and potentially disabling discomfort that might follow from really 'thinking globally'. Because if I *really* did that, as matters have now come to stand, I might despair.

Optimism as denial

What we can see emerging here is a sub-variety that we might call *self-motivating* or *activist* denial. This is denial customised for people who want to get something done about our environmental plight. It involves the willed ignoring of what might inhibit the energy needed to motivate would-be remedial action, for which it must always therefore be assumed that there is still (just) time. The gravity of the looming danger is acknowledged; so is the way it calls for, not only fully justified anxiety, but decisive lifestyle change. What is downplayed is the implication, following from a full recognition of its seriousness, that the danger cannot, after all, now be averted.

But implicatory denial, while it is possible, is always going to be unstable for the environmental activist, whose *raison d'être* as an activist is precisely *not* to change the subject. On the other hand, being always uneasily aware of the Vicious Syllogism, the serious activist cannot simply refuse to accept (in an inversion of literal denial) that it is too late. The only place to go is therefore willed optimism: *interpretive* denial for activists. Essentially, here, the newly closed window of opportunity is reinterpreted as still open. I want to illustrate this by addressing a couple of recent discussions of climate change where this notion does seem to help explain what are otherwise quite startling disjunctures of thought and argument.

Since these analyses will necessarily be critical, I shall start, albeit briefly, with self-criticism. Denial of this kind is what, I now recognise, spoilt the final third of my own last book, *The Sustainability Mirage*.[9] Having shown what was wrong with the sustainable development model for environmental policy, and having sketched very inadequately a different approach to thinking about ourselves in relation to the future, I still felt that I had to show how this kind of re-conception might give us a last chance of keeping carbon emissions below the danger threshold. Hence, I invested heavily in Contraction and Convergence, the idea that the world's economies negotiate carbon budgets for their citizens which gradually approximate, over an agreed timescale, until everyone on Earth has the same carbon allocation at a level with which the atmosphere can cope – so simple a solution, so equitable, so helplessly utopian. Writing in 2006–07, when Blair's New Labour government was dying on its feet and its commitments to sustainability had been revealed to be as vacuous as everything else it said, I even managed to make myself think that a shift towards a more entrepreneurial model under the Conservatives might contribute towards the kind of localising self-reliance on which such a global carbon regime would ultimately depend. That is a very bitter recollection now, in view of the destructive and utterly irresponsible performance of the post-2010 coalition – the self-professed 'greenest government ever' – once actually in office; but willed optimism looking for an interpretation of events on which to base itself does readily lend itself to being taken in like that. I remained, in short, captive to the paradigm of progress, the idea that our continued development, *somehow*, towards a better future simply cannot be questioned.

My sketch of possibilities nevertheless rang hollow, being actually a willed refusal of what I had already argued about the pseudo-constraints and operational helplessness of sustainable development policy. I was engaged, precisely, in a form of interpretive denial. The policy goal – avoidance of catastrophic climate disturbance – was unachievable on the sustainable development paradigm, because like a mirage that paradigm had failure to get there built into it. But reinterpreted as 'deep sustainability' – essentially, an attitude of adaptive responsiveness to emergent events – our urge to save ourselves for continued progress could still be represented as having some prospect of success. The mirage *could* be reached, after all.

Willed optimism, that is, warps thought – and evidently, not just other people's. I now turn, however, to a much better-known example, Mark Lynas' most recent book *The God Species*, which shows where self-motivating denial can take an established and well-respected environmental commentator.

Lynas is an excellent science journalist, as well as a committed and intelligent environmentalist. I have already mentioned his book *Six Degrees: Our Future on a Hotter Planet*, which assembled with formidable thoroughness the main scientifically based predictions for what the world would be like under global average temperature rises of 2 °C, 3 °C … and on to 6 °C. This book is original, as well as extraordinarily useful, in the meticulous clarity with which the scenarios for different areas, the interactive ecological dynamics and the grim causal logic of the process are detailed. In its final chapter, it presents with characteristic force what is nevertheless a familiar green response to this challenge: we need to change our lifestyles dramatically and quickly, but with effort and moral energy assisted by strict statutory carbon rationing, this could still be achieved.

We can cut our need for energy by leading less consumptive lifestyles and by adopting more localised patterns of behaviour ... All the evidence shows that people who do not drive, do not fly on planes, do shop locally, do grow their own food ... have a much higher quality of life than their compatriots who still persist in making the ultimate sacrifice of wasting their lives commuting to work in cars ... In constraining carbon through rationing, we might soon find that we were building a different sort of society, one emphasising quality of life before the raw statistics of economic growth and relentless consumerism.[10]

But that was then. Four years on, we have *The God Species*. In between, of course, had come Copenhagen. Unsurprisingly, this seems to have affected Lynas, too, as a moment of truth. But the truth which it revealed was not, apparently, the bankruptcy of environmental optimism, but the increasingly desperate need to reassert and reinforce it. Humans can only get out of this mess, it would now seem, by quite explicitly embracing the God-role that evolution has thrust upon us, and deliberately undertaking to manage the planet by technological means.

The new book sets out this managerial agenda in terms of nine 'planetary boundaries', a concept taken up from the work of a team of scientists headed by Johan Rockstrom of the Stockholm Resilience Centre.[11] The idea is essentially an extension of the tipping-point approach to global warming. In this and eight other areas (biodiversity loss, nitrogen fixation, land use change, freshwater usage, toxic chemical pollution, atmospheric aerosol loading, ocean acidification and stratospheric ozone depletion), threshold values of one key parameter in each case either have been or are in the process of being identified. These quantify as precisely as possible the points beyond which the given kinds of pressure on the respective environmental media exceed natural capacities for resilience and initiate runaway deterioration. Keeping below these thresholds (or, in the case of the first three, getting quickly back below them) would mean that humanity remained within the boundaries of the overall 'safe operating space' provided by the Earth for pursuit of our long-term social and economic development goals.

As such, this is an ambitious attempt to add some clear and simple hard-number targets to the basic sustainability model, so that these can serve to license the battery of technological fixes now needed to deliver it. But what is really striking is the amount of traditional green cargo that Lynas is now happy to see ditched in order to keep the ship afloat. Thus, opposition to nuclear power and GM crops is dismissed as 'Luddite prejudice'. Restraining economic growth is mistaken because 'increasing prosperity – measured in material consumption – is non-negotiable both politically and socially'.[12]

Persuading people to give up flying is a non-starter (as well as hypocritical since most environmentalists still fly). Jonathon Porritt is reprimanded for insulting people by suggesting that they might make reproductive choices in the light of a concern for global over-population. The New Economics Foundation is mocked for claiming that people in some poorer countries are happier than the typical resident of megalopolis. Urbanisation is in any case a good thing, since cities not only offer people more opportunities but give biodiversity a chance to re-establish itself in the depopulated countryside, while agribusinesses deploying genetically modified nitrogen-fixing cereal crops get on with producing food for nine billion people (and rising) far more efficiently than organics could ever hope to achieve.

This chivvying of sacred cows overboard to the sharks is certainly fun to watch. But the vessel left buoyant after the supposedly dispensable cargo has been dispensed with comes down to global capitalism *faute de mieux*, market incentives, the pricing of key ecological services and an armoury of high-tech interventions, all managed by an internationally collaborating humanity stepping up cheerfully to its new God-like planetary management duties. Given where Lynas was coming from only four years earlier, this represents a very dramatic shift of position indeed.

In pointing this out, I am certainly not berating him for changing his mind – indeed, given what I have just been saying about myself, how could I? (As I recall to have learned from one of the *Dr Dolittle* books a very long time ago, if you ever change your underwear, you should always be open to the possibility of at least sometimes changing your mind.) But there are definite signs that the shift in Lynas' case owes less to new insights in the meanwhile, and much more to the intensifying pressure of willed optimism, than appears on the surface. Willed optimism, I have suggested, represents an intelligent activist's form of interpretive denial. As such, it is going to operate under a good deal of internal strain; and for all its scientific literacy, there are arguments at key junctures of *The God Species* which suggest exactly that kind of strain.

For instance, crucial to the whole 'God-species' move and concept is this claim, which comes early on:

> Most Greens ... emphatically object to geo-engineering – the idea that we could consciously alter the atmosphere to counteract climate change, for example by spraying sulphates high in the stratosphere to act as a sunscreen. But the objectors seem to forget that we are already carrying out massive geo-engineering every day, as a hundred million people step into their cars, a billion farmers dig their ploughs into the soil, and ten million fishermen cast their nets.[13]

But of course, the geo-engineering that might be planned in government offices is a very different thing from the geo-engineering achieved by the 'invisible elbow' of these multiple uncoordinated economic activities. (I first heard this nice term for externalities from the environmental economist Michael Jacobs: attached to Adam Smith's 'invisible hand', by which capitalism supposedly produces its beneficial consequences, there is the invisible elbow with which it at the same time unintentionally knocks things over.[14]) Now when one actually thinks about it, the difference between planned and elbowed geo-engineering is pretty clearly so great that to make anything turn on applying the same term to both is a quite illegitimate procedure. Lynas seems at one level uneasily aware of this – elsewhere he tries to fudge the point by talking about 'accidental' as opposed to 'deliberate' planetary management,[15] as though 'accidental management' were not a contradiction in terms. But his unease doesn't stop him implying that because we are accustomed to unplanned 'geo-engineering' we are being *inconsistent* (rather than, say, pragmatically fearful of human over-reaching, or just plain realistic about the likely reach of Sod's Law) in objecting to the planned variety – or, as he puts it even more clearly in a book-launch speech available on his website: '*not* geo-engineering is not an option in actual fact because we are already doing it'.[16]

This, however, is just to argue by equivocation, the fallacy of using the same word with significantly different meanings in different premises and then claiming that *any* conclusion follows. At issue here is not, I should emphasise, the feasibility (or not) of various technological fixes – and of course, a claim is not falsified because someone offers a bad argument for it. My point in this context is the *glaring* badness of the argument, which surely couldn't have been overlooked by anyone who wasn't determined not to see it. In order to present God-fixes as even *prima facie* feasible, even so acute a commentator as Lynas can be seen to be relying on what is, in fact, a crude sleight-of-hand for not recognising what is under his nose. Willed optimism, to repeat, warps thought – indeed, can only be held in place by warping thought.

Again, we are urged to disembarrass ourselves of unhelpful Green prejudice in these terms:

> Global warming is not about overconsumption, morality, ideology or capitalism. It is largely the result of human beings generating energy by burning hydrocarbons and coal. It is, in other words, a technical problem, and it is therefore amenable to a largely technical solution.[17]

But this – and even more glaringly – is mere *non sequitur*, the confident production of a conclusion that simply doesn't follow from the given premises.
It would only have a chance of following if humans burning hydrocarbons,
on the one hand, and overconsumption or capitalism on the other, were
alternative explanations for global warming, which very clearly they are not.
Much more plausible is to say that the burning is essential to the overconsumption which is driven by the capitalism, and if that is the dynamic it
may well make any attempt at a purely technical solution to the burning
unlikely to stick for long. The effect of technicist tough-mindedness is only
achieved here by another sleight-of-hand, trying to palm away this very
obvious response. The trouble is, again, that the sleight-of-hand is itself so
obvious that it would surely not have satisfied anyone who wasn't determined not to see it.

The God Species is an extremely interesting and useful book. Like all
Lynas' work, it is not only scientifically well informed, but possessed of a
verve and clarity in the writing which make scientific concepts and technological possibilities more accessible to the lay person than does any other
contemporary commentator on these issues whom I've come across. It is
precisely because of these virtues that I have picked on it to represent a
trend that Paul Kingsnorth has labelled 'neo-environmentalist thinking'.
This he characterises as

> a progressive, business-friendly, postmodern take on the environmen-
> tal dilemma. It dismisses traditional green thinking, with its emphasis
> on limits and transforming societal values, as naive. New technolo-
> gies, global capitalism and western-style development are not the
> problem but the solution. The future lies in enthusiastically embrac-
> ing biotechnology, synthetic biology, nuclear power, nanotechnology,
> geo-engineering and anything else new and complex …[18]

Key figures in this movement of thought include the 'eco-pragmatist'
Stewart Brand ('We are as gods and have to get good at it'), the American
conservation writer Emma Marris and a group of others centred on organisations such as the Breakthrough Institute and the Copenhagen Consensus.
The strident unrealism which all these voices endorse (the Breakthrough
Institute's mission, for example, is 'to accelerate the transition to a future
where all the world's inhabitants can enjoy secure, free, and prosperous lives
on an ecologically vibrant planet' – no Vicious Syllogisms here, clearly) is
courageously and openly carried through to its techno-logical conclusion
in Lynas' book. In the process, I suggest, it reveals itself clearly as a form
of interpretive denial. What its willed optimism commits it to reinterpret,

we might say (casting back to the Prologue), is the tragedy of our situation, which it still reads as a set of problems – extraordinarily demanding ones now, but still by definition susceptible of solutions, if only we are bold enough to grasp them.

If this is plausible, we can characterise the emerging crisis within environmentalism as a sense, no longer capable of being suppressed, of the strains involved in this kind of denial. On the one hand, environmentalism's own arguments lead convincingly, via the Vicious Syllogism, to the belief that it is now too late to prevent severe anthropogenic climate disruption and the associated worldwide ecological damage. On the other hand, it seems that activists must go on interpreting the situation and its possibilities so that it is still, despite all the appearances, *not* too late. The force of this obligation is made very clear by Lynas: humans have to assume the God-role to save the planet because failure is inconceivable, and therefore we must be able to succeed. To believe anything less is pessimism (which must be what you get if you abandon optimism), and – to repeat what I have already quoted from him – 'only optimism can give us the motivation and passion we will need to succeed'. The connection between optimism and action asserted here is crucial. To act constructively requires a motivating energy of which despair, the abandonment of hope, will rob us – and the non-abandonment of hope is optimism, the habit of expectation that all will somehow be well. The perception of its being too late is therefore denied, although because it is so well grounded in the best environmental information and analysis, these grounds must also at the same time be acknowledged. The resultant denial remains seriously unstable: recognising that it is too late, we refuse to give up acting as though it wasn't, willing ourselves to insist on what we know (and not far beneath the surface of our minds) to be incredible.

The culture of denial: sustainable development

We have now considered both community routines for normalising evasion and the fixed-grin optimism of the neo-environmentalists as forms of environmental denial in practice. But to identify the pervasive culture of denial within which these and similar practices subsist, and the context of assumptions that licenses such denial, we must turn to the sustainable development paradigm.

This can be seen as a conceptual framework designed to legitimate the illusions that we have seen being given comfort in these different forms of denial. Or, if *designed* implausibly suggests deliberate intention, let us say: *socially constructed*. This may serve to recall the 'social production of innocence' identified by Norgaard in her Norwegian case study – and very appropriately, since I believe that the sustainable development paradigm and its widespread adoption as the politically acceptable face of environmentalism over the past thirty years together represent a complexity of social process aimed at achieving just that kind of tacit union of acknowledgement with exculpation. Sustainable development is a hook that is meant to let you off it, an unconstraining constraint, an acceptance of obligation so configured as to excuse us from being obliged.

Emphasising this aspect also helps me with a problem of presentation. I have discussed sustainable development in print before, and inevitably (given my concerns) from the same perspective. If only for reasons of space, I don't want to repeat all of what I said about it in the first sixty-five pages of *The Sustainability Mirage*, despite continuing to stand firmly by that analysis. But nor can I assume that all present readers will have that book to hand. I have really no option therefore but to summarise pretty drastically the account which I have already published, with its earlier outing explicitly in the background for anyone interested in a fuller treatment. But presenting this account in the context of an analysis of denial may shed helpful new light.

According to the sustainable development picture, what it means for our actions and refrainings in the present to matter environmentally is that they have in-principle measurable future consequences for in-principle identifiable ecological resources supporting economic activity. These consequences are positive or negative, to be pursued or avoided, insofar as they either contribute to or derogate from the ability of people in the future to go on living at welfare levels at least equivalent to our own. Sustainable development thus has the appearance of being a hard-headed, operationalisable way of behaving justly towards the future by making changes in our present activities and lifestyles. It looks operationalisable because having quantified what the future will need by way of ecological resources for maintaining welfare levels, we can then work back, again quantifiably, to what we must do or refrain from doing now to ensure that those future needs can be met.

But this whole future-focussed, would-be scientific model of environmental concern isn't actually part of the solution – rather, it is a deeply embedded part of the problem. That is because its purported commitment to the future is as full of escape clauses as a dodgy insurance contract. This, in turn, is because it depends on gross scientistic exaggeration of our powers to predict and control, and does so in the service of supposed obligations to future people which actually are no more than pseudo-obligations. Hence its appearance of being operationalisable comes down to its supplying us with a battery of *floating standards* – that is, sustainability benchmarks and targets that flex according to how comfortable or uncomfortable we find it to be constrained by them.

In an alternative image, we can think of sustainability policy and planning as like trying to mend your old rusty bike with a set of lead spanners: under pressure, what is always going to give is the spanner, not the nut. Sustainability thinking involves putting supposedly plausible numbers on the future in order to gain leverage for present socio-economic and lifestyle changes that we find eminently resistible. Entirely unsurprisingly, therefore, the numbers gain no real purchase. Either they are methodologically manipulated, as in much contingent valuation for instance, to produce the result we first thought of, or they are presented as duly drastic and then ignored – insofar as treating them as not *robustly* drastic comes to the same thing as ignoring them. This is possible because we are tacitly aware all along that the figures *aren't* robust, that they are essentially manufactured, that there is no way of genuinely quantifying the effects of specific present actions or refrainings on the mid-term future parameters of almost unimaginably complex ecological interconnectedness under endemic uncertainty, without making huge assumptions from which the numbers inherit enough contingency to be little more than gestural.

Underlying these features of the sustainable development policy model is something that we are constantly tempted to forget: that the future is always, and only ever, *under construction* in the immediate present. There is nothing existent beyond the advancing edge of the present – the future exists just in virtue of the fact that the edge *is* always advancing. Moving into it, in other words, is really *nothing* like negotiating a terrain with an established topography. Instead, in the words of the sociologist John Urry, it is like 'walking through a maze whose walls rearrange themselves as one walks through: new footsteps have to be taken in order to adjust to the walls of the maze that are adapting to each movement made ...'.[19] The future is always under construction to a plan that is itself being perpetually reconstructed.

We also, it is true, reconstruct the past, in memory and imagination – we have no other cognitive access to it, after all – but the radical difference is that these constructions are made true or false (whether we know it or not) by what did actually happen. Thus, they do to some extent parallel the way in which a map can represent, more or less accurately, a mapped terrain. The distinction between the actual past and our construction of it is then one that we can put to good use – in trying to confirm our account of past events by finding out more about what *really* happened, for instance. However, with the distinction between what will happen and what we think will happen – and even more crucially, with that between what we think will happen and what we *want* to think will happen – there is no equivalent point of external reference. Nor, by the same token, is there a firm enough distinction for any genuinely constraining obligation to get a grip, between what we suppose that as yet non-existent future people will legitimately expect us to have handed on to them, and what we want to suppose is due to them.

Thinking about the future in this misleading way is, as I say, inherent in scientism – that is, the very general contemporary disposition to expect more from science than it could possibly deliver. In practice, we live with the uncertainty and malleability of the future all the time – it is the permanent sub-text of human agency, and anyone who has ever seen his week's carefully planned schedule go pear-shaped is a folk sociologist of scientific knowledge. However, embedded in the formation of sustainable development is the idea that 'rational' science-based planning can somehow transcend these inherent conditions on our knowledge and action. Hence, 'strategies' envisage specific percentage cuts in CO_2 emissions by specific dates fifty (or twenty) years ahead, as a way of keeping below specific global warming thresholds, and assume that adopting such a planning horizon is an act of sober and responsible administrative realism – when what it really represents is hubristic disdain for the glaring fact that all the ecological synergies and feedbacks involved are infinitely too complex to guess at beyond the very short term, never mind predict and try to plan for.

Licensing denial

The American historian of science Theodore Porter provides a brilliant cameo case study that I mentioned in *The Sustainability Mirage* but can't forbear to quote again at this point, because it encapsulates so neatly what is actually at issue here:

> A congressional mandate permits the United States Forest Service to cut no more lumber than is renewed by annual growth. Since that law was put into effect, growth rates have been greatly enhanced, at least in the Forest Service accounts, by new herbicides, pesticides and tree varieties.[20]

Here we can see very clearly in miniature just how the relevant self-deception works. The inadmissible real dilemma – 'We have to appear to be complying with the mandate, but we really do need the present income from going on cutting the lumber' – becomes the admissible ('sustainability') aspiration to boost compensatory growth rates well above previous levels. This in turn depends on finding means to quantify the forecast effects of new interventions in complex living systems so that they show this to be what, over the longer term, is actually being done. An underlying awareness both of the need for accountability and of the scope of indeterminacy allows the numbers for growth rates to be regarded as firm in the accounts presented to Congress, while being recognised as suitably soft in the process of estimating them from the projected effects of new treatments and tree varieties. A constraint has been evaded in bad faith, as Porter's quiet irony intimates, but nobody at any stage has had to lie!

Taking the hint from this, we can construct a similar characterisation of what is going on in the kind of activist denial exemplified by Lynas and his coadjutors. We have recognised the now very pressing urgency, we might say, of being in time to avert climate catastrophe. Since that recognition, our capacities to keep within the threshold values of the key planetary boundaries have been greatly enhanced, at least in our reports, books, policy documents and funding applications, by a variety of new technological and geo-engineering possibilities. (The reader can very easily complete the parallel.)

Now Cohen defined implicatory denial as 'not a refusal to acknowledge reality, but a denial of its significance or implications'.[21] Doing this at the individual level frequently depends, he notes, on a 'rich, convoluted and ever-increasing vocabulary for bridging the moral and psychic gap between what you know and what you do'. We might characterise the use of such a vocabulary as a discourse within which recognition of the intolerable as intolerable can coexist with its toleration. Norgaard's case study showed this kind of thing happening in relation to global warming: the use of a multilevelled, partly generalised and in part locally specific discourse permitting acceptance of responsibility for climate change to cohabit consciously with the lifestyles that continue, and are recognised as continuing, to drive it. What should now be clear is that the sustainable development paradigm offers a discourse on a larger sociopolitical scale and in a quasi-scientific mode, with a precisely parallel structure and function: a discourse within which we can both recognise unevadable environmental constraint in all its forms, and still actually evade it.

It must be stressed that a key function of denial is not just to *permit* toleration of the intolerable, simply to look the other way while we sneakily let ourselves off doing what we don't want to watch ourselves failing to do. It is, even more dangerously, to appear to *license* that process, through a denial of itself as denial. So sustainable development licenses our escape from taking our environmental obligations too uncomfortably seriously by giving us extensive science-based sanction for holding that taking them seriously is just what we are doing. It does this, in fact, by embodying both interpretive and implicatory denial, and enabling the tacit move back and forth between them as required. Thus we treat the pseudo-numbers on which it trades as providing firm operationalisable targets when we want to interpret what we are confronted with only as soluble problems, about which we are hard-headedly doing something. But we can also treat them as flexible guesstimates when we want to evade the implications, that is, to not actually *do* whatever is too uncomfortably required of us.

This is an adaptable conceptual tool, and it can perform its function in different ways for different phases of the struggle. Thus it serves its originally intended purpose of reassuring us that we can indeed go on with 'development' (conceived as material progress in the broad sense), so long as it is kept 'sustainable'. This is only a matter of keeping within the 'planetary boundaries', to recur to Lynas' formulation of an updated sustainable development model – and that these boundaries are quantifiable is reassuring even when (as with climate change) we have already transgressed them: at least we have a clear-cut task of recovering well-defined ground. That the ground is not really well defined, and is tacitly known not to be, at the same time prevents the challenge from appearing too daunting. The numbers are never really determinate enough to substantiate the charge that to have gone only *thus* far, rather than somewhat further, wasn't meeting our obligations – and, knowing this, we can settle comfortably in practice for roughly what we find we can manage. That is why, for instance, the UK Labour government which introduced the most (theoretically) stringent set of public sustainable development commitments in the world,[22] commitments that *blatantly* required, among other things, a massive reduction in recreational air travel, could at the same time plan to build yet another runway at Heathrow. The sustainable development model of environmental responsibility will spur us to go just far enough not to make any difference that hurts.

But if it is a matter, further down the line, of reassuring ourselves that there is still time to avoid catastrophe, the model serves equally well. Relying on pretend-determinations of what is at the same time tacitly acknowledged to be indeterminable means that no failure to do enough for the habitability of the future by way of present adaptation or mitigation is ever going to be sufficiently decisive *now* to prevent us from incurring it, accommodating it, cheering ourselves up and moving purposefully onwards. This allows for all the elasticity of timescales and adjustability of deadlines that we need in order to enable us – like Christiana Figueres, like so many since Copenhagen, since Cancun, since Rio+20 … – to go on working and hoping optimistically for the best.

Denial and progress

Why have we constructed, collectively, these practices and institutions of denial?

We shall not answer that question satisfactorily, in my view, by appealing as the psychoanalyst Sally Weintrobe does in her recent collection to self-defence against environmental anxiety – protecting ourselves by disavowal against what she calls 'an environmental neurosis, rooted in deep-seated annihilation anxiety'.[23] Undoubtedly something of that kind is likely to be contributing to the state of mind of those whose involvement in climate change denial is a way of trying to reassure themselves that the problems of which they are uneasily aware aren't so serious, or at least aren't intractable. But I want to lay great stress at this juncture on one of the main observations of this chapter, which is also a key premise for the whole book. That is that denial is very plainly confined neither (to repeat) to the bad guys who just want to go on coining money while they can, nor to this large and still growing class of what we might call the nervously wishful good – the people who need (as I noted in the Prologue) to shunt what they recognise or suspect is happening to the backs of their minds because they have lives to lead, jobs to do, books to write and children to raise under the conditions in which they presently find themselves. The characteristic structures and practices of denial are also fully exhibited, as I hope to have shown, by environmental activists: by people whose main or at any rate salient life-purposes involve full explicit acknowledgement that the threats posed by climate change are real, unignorable and increasingly imminent.

People with these concerns and goals do not want to disavow the anxiety-producing aspects of these threats – on the contrary, they want to flag them up, to feed their projects and inform their messages. They may, indeed, be made anxious by the thought that their activity up to this point has been in vain, and they haven't managed to prevent what they have been campaigning against. Environmental campaigning of any sort still takes a huge amount of commitment, and such a reaction would be perfectly natural. But it would be peculiarly self-defeating to defend themselves against it by pretending that things aren't really so bad. And yet we have found serious activists implicated in forms of denial that are very clearly continuous with those out of which they want to shock others, and sharing almost without exception in the basic assumptions of the culture of denial and bad faith which is sustainable development.

An historical analogy can perhaps help to clarify matters here. Reluctance to face up to uncomfortable facts that are at the same time half-recognised to be true is of course a very long-standing human characteristic; we all know individuals who act in this way, and history shows us many examples of societies exhibiting the trait collectively. A fairly recent classic case was 'Appeasement', the resolve in British official circles during the 1930s, reflecting a widespread feeling in society at large, not to recognise for what it was the programme of armed aggression being prepared by the totalitarian regimes of Mussolini and Hitler. There is a passage in Churchill's memoirs of this period dealing with an early manifestation of the approach, in response to the Italian invasion of Abyssinia in 1935, which bears quoting at length because of the gleaming sidelight that it throws on more modern affairs:

> The Prime Minister [*Baldwin*] had declared that Sanctions [*against Italy*] meant war; secondly, he was resolved there must be no war; and, thirdly, he decided upon Sanctions. It was evidently impossible to reconcile these three conditions. Under the guidance of Britain and the pressures of Laval [*then French Foreign Minister*] the League of Nations Committee charged with devising Sanctions, kept clear of any that would provoke war. A large number of commodities, some of which were war materials, were prohibited from entering Italy, and an imposing schedule was drawn up. But oil, without which the campaign in Abyssinia could not have been maintained, continued to enter freely, because it was understood that to stop it meant war ... The export of aluminium to Italy was strictly forbidden; but aluminium was almost the only metal that Italy produced in quantities beyond her own need. The importation of scrap iron and iron ore into Italy was sternly vetoed in the name of public justice. But the Italian metallurgical industry made but little use of them, and as steel billets and pig iron were not interfered with, Italy suffered no hindrance. Thus the measures pressed with so great a parade were not real sanctions to paralyse the aggressor, but merely such half-hearted sanctions as the aggressor would tolerate ... The League of Nations therefore proceeded to the rescue of Abyssinia on the basis that nothing must be done to hamper the invading Italian armies.[24]

The parallels here with contemporary international action over climate change are painfully obvious. The principal function of mainstream sustainable development, it becomes clear, is in fact to license the Appeasement of our time. The world, represented now by the UN as successor to the inter-war League of Nations, has recognised that seriously reducing greenhouse gas emissions means stringently restricting further fossil fuel usage; second, it is resolved (tacitly, but no less firmly for that) on there being no such stringency; and third, it has committed itself, over and over again, to seriously reducing greenhouse gas emissions. Once more I can leave it to the reader to complete the parallel between deliberately ineffectual sanctions and the floating standards and ever-receding targets of climate policy under the sustainable development approach. It is intriguing, too, to identify the equivalents of our current deniers in the various strands of opinion that supported Thirties appeasement as a policy: the literal (Hitler and Mussolini represented no threat, as Rothermere's *Daily Mail* regularly announced); the interpretive (their demands could be seen as reasonable, and so should be placated, according to Geoffrey Dawson's *Times* editorials); and the implicatory (what after all was Czechoslovakia, as Chamberlain notoriously said at the time of Munich, but 'a far-away country' containing 'people of whom we know nothing'?)

But then, where were the activist deniers – the equivalents of those who now campaign for the compelling arguments enforcing the Vicious Syllogism, but still continue to insist on looking for implausible wiggle-room over climate change? This is where the parallel revealingly breaks down. There was, indeed, no one around in the 1930s who said: 'Now that a rearmed Germany is led by Hitler, another war with her is almost inevitable – so rather than vigorously rearming ourselves, we must strive ever harder to avoid war.' Such a position would have been just too obviously untenable. Those comparatively few such as Churchill himself who recognised the real situation, recognised also that the only response was urgent war preparation. But then, the anti-appeasers had no equivalent of sustainable development thinking available to them – no sanctioned and well-regarded way of setting what they knew to be an irremediably soft target for aircraft production, for instance, while reassuring themselves because (on paper) they had the Luftwaffe beaten. The untenability of activist denial over climate change, in other words, is only possible because its central inconsistent move – invoking the inevitable as a desperate spur to avoidance – can be blurred from view by the dominance of the sustainable development paradigm with its inherent quantitative slipperiness and aspirational bad faith.

But what pressure could be powerful enough to push people into such an untenable position *at all*? We shall not explain this without digging down to a level at which we can find the possibility of tragedy, rather than any form of neurosis. Willed environmental optimism, I want to argue, is an aspect or manifestation of the same destructive dynamic against which it superficially sets itself. It is thoroughly implicated in the basic commitment of the modern world that drives the destruction. That is, in shorthand, the commitment to *progress*, the continuous overall improvement of the human condition. It is a commitment that the modern world has hitherto seemed quite unable to question, although in the various forms of denial associated with the environmental challenge this inability has now taken a clearly pathological turn. Environmentalism certainly starts from the recognition that progress as crudely defined by the mid-twentieth century, the raising of as great a proportion as possible of (potentially) nine billion human beings to something like the carbon-based material standard of living prevailing in the advanced North and West of the world, is simply beyond the ecological resources of the planet. But the mainstream interpretation of environmentalism as 'sustainable development' then turns that into a commitment to the continuation of progress by other means. If the betterment of the currently 'under-developed', and the yet further betterment of the already 'developed', cannot be pursued indefinitely through technologies and economic arrangements dependent on the burning of hydrocarbons and coal, then technologies and economic arrangements must be found on the basis of which some version of it *can* be pursued indefinitely (or 'sustainably'). There is a collective, socially embedded refusal to accept, whatever the evidence, that permanent progress, permanent improvement in the standards and experienced quality of as many human lives as possible, can ever cease to be a practicable goal. The various kinds of denial then attach themselves to different aspects of this refusal.

And why does the modern world have to resort to such accumulating denials – straining credulity as, viewed dispassionately, they clearly do? Why can't we shake free of that commitment to progress? Why are we locked into the fantasy of perpetual cumulative improvement on a now evidently finite planet? The answer that I want to explore is that human beings have to find life liveable in a way that is not subverted by the knowledge, from within, of our own mortality. But for a complex of historical and cultural reasons, we in the dominant secular Western civilisation have now no conceptual resources for expressing life-purpose from this internal perspective except in terms of *improving the future* – bettering the conditions of life, first for ourselves (extending to the whole population of the globe) and then for those who come after (extending indefinitely into the future). This is the 'social hope' that the egregious C.P. Snow pretended to offer, in a once-famous disquisition,[25] as compensation for our all having to die individually. Here the necessary analysis is existential rather than psychoanalytic – it is to do with the conditions imposed on us by conscious being as such. The commitment to progress is at bottom a doomed attempt at running away from an inability to understand ourselves as naturally finite beings, which in turn arises from a deep disjunction between our lives as we now live them and the structural needs of reflexive organic consciousness.

Here in these last paragraphs, I have outlined the agenda for investigating the roots of climate change denial – in all its manifestations – which I shall be pursuing through the next two chapters, and then on into the heart of the book.

Notes

1 See Cohen (2001).
2 In Sartre (1943/1958), especially Chapter 2 (the quotation is from p. 56).
3 See his papers 'On Negation' (1925) and 'Fetishism' (1927) in Freud (1961), vols. XIX, pp. 235–9 and XXI, pp. 149–57 respectively.
4 See Foster (2008) and Weintrobe (2013), respectively.
5 Cohen (2001), p. 7.
6 Norgaard (2011).
7 Norgaard (2011), p. 97.
8 Norgaard (2011), p. 83.
9 Foster (2008).
10 Lynas (2007), pp. 299–300.
11 Lynas (2011), pp. 8–9.
12 Lynas (2011), p. 67.
13 Lynas (2011), p. 11.
14 See Jacobs (1991), p. 127.
15 Lynas (2011), p. 196.
16 See www.marklynas.org/2011/07/the-need-for-a-rational-environmentalism-speech-at-god-species-launch/
17 Lynas (2011), p. 66.
18 See Paul Kingsnorth, 'Dark Ecology', published in *Orion / Dark Mountain*, January 2013, also available at: www.paulkingsnorth.net/journalism/dark-ecology/
19 See Urry (2005).
20 See Porter (1995), p. 44.
21 Cohen (2001), p. 8.
22 The Climate Change Act 2008 c. 27 (London: HMSO).
23 Weintrobe (2013), p. 41.
24 Churchill (1948), pp. 157–8.
25 Snow (1959), p. 9.

2
PROGRESS PAST AND PRESENT

I suggested at the end of Chapter 1 that the present crisis of environmentalism, manifesting itself in various related forms of willed denial, reflects a deeply ambiguous attitude towards the concept and practice of progress – on the one hand, ecological critique, on the other, a pathological form of commitment. In this coming chapter, I want to trace how that ambiguity developed and took hold as the environmental movement emerged in response to the hazarding of planetary life which the pursuit of progress was, by the mid-twentieth century, evidently bringing in its train. Specifically, I shall try to show how the emergent environmental movement compromised itself as an epochal critique of that pursuit, through failing to follow through on its recognition of the crisis of spirit which that crisis of resources also, and inseparably, represented. Through this account, we can perhaps begin to appreciate some of the reasons why activist denial may now be so deeply embedded in the sustainable development culture.

It will be helpful in the first place to cast back briefly, and consider in outline where the idea of progress originally came from, what its dimensions are, how it has realised itself in material practice over time, and how it has been closely interwoven with other-than-material forms of human aspiration. I will then go on to look at the relation of environmentalism to this history.

The progress of 'progress'

The American historian Robert Nisbet, whom I have already quoted, identifies the fundamental notion of progress as the idea that 'mankind has advanced in the past – from some aboriginal condition of primitive-ness, barbarism or even nullity – is now advancing, and will continue to advance through the foreseeable future'.[1] As Nisbet shows with a wealth of scholarly detail, this idea has well-established Classical antecedents, being identifiable even in Homer – though it coexisted throughout the Classical period with the alternative thought that degeneration rather than progress offered the truer picture of human history. The idea of progress as perpetual advance then informed the Christian conception of history from patristic times onwards, survived the obscurantism of the mediaeval period and the Machiavellian realpolitik of the Renaissance, and with the Reformation took particularly vivid forms in the Protestant millenarianism and associ-ated political radicalism of the seventeenth century. But the form in which it had come by the twentieth century to represent what Christopher Lasch calls the working faith of our civilisation[2] – that civilisation whose key assumptions about economic growth as the engine of progress the environ-mental movement arose in large part to challenge – was secular, technicist and liberal-democratic.

It was essentially secular because it was and is primarily concerned with what may be called the external, material conditions of life. It is important to be clear what is involved here. It is not just a matter of the progres-sive improvement of human well-being conceived as the acquisition of more *things* – more and better fridges, cars, mobile phones, solar panels …, year on year. That is just a parody, though an understandable one given how central the advertising of consumer goods is to capitalism's domi-nant version of the idea of progress. The concept of material progress can accommodate a wider view of human well-being than the steady accu-mulation of such goods; it remains a fundamentally materialist conception nevertheless.

Improving well-being, on this view, is all about progressively diminish-ing the pains and frustrations imposed on human beings by their situatedness in a world of experience structured by time and space. We are more acutely conscious of our desires than other sentient creatures, because we do not just feel them, but are capable of conceptualising and appreciating the tem-poral and spatial distances between them and their satisfaction. We don't just want what we want, we recognise and want the means to what we want, and suffer consciously from the lack of them. So while improv-ing our well-being can indeed be about increasing our access to ordinary, direct satisfiers such as food, warmth and sex, it is also very much about enabling us, once these basic needs are met, to access more quickly and eas-ily, or over greater physical distances, other things we want, many of which may not be obviously material at all, but the objects of love or pilgrimage or aesthetic delight or intellectual fascination. These are traditionally thought of as things of the spirit, and any notion of 'progressive improvement' seems to travesty our relation to them. But we always encounter them in material settings about which such a notion makes perfectly good sense. To take a very simple instance, gaining internet access to the library catalogue, rather than having to walk to the library, constitutes *material progress* in this sense (so long as one can still get the exercise from walking in other ways) – although how much I get out of the books which I then read will require to be assessed in a different dimension. Ditto with getting to Birmingham twenty minutes sooner on the proposed high-speed train, though it will still be Birmingham that you have got to. At another level, the state of my soul cannot be measured, while how much I have to eat can be – but I will have little time to attend to the state of my soul until I have at least enough to eat. Material considerations matter because we are material beings as a condition of being anything else besides, and so improvement in our material situation is always to that extent genuine advance.

Also inevitably involved in the idea of material improvement thus understood is the extent to which its benefits can be distributed equitably and reliably – since well-being includes our not being unduly disadvantaged in relation to others in these respects, and as the species that has evolved ethics we are also capable of appreciating the corollary, that others should equally not be unduly disadvantaged in relation to ourselves. So the demands of justice, and of its necessary conditions (law, education, peace ...) all feature too in what nevertheless remains an essentially material understanding of human progress.

Why this understanding of progress in fundamentally material terms should have come so to predominate, when in its Baconian and Enlightenment origins the concept had at least as much to do with the emergence of the human spirit from bondage to ignorance and superstition, is a key question. Much of the answer has to do with the second characteristic of progress as manifested in the early and developing modern age – its relation to cumulative advances in science and technology. Recognition of this dynamic connection goes back at least as far as Thomas Sprat's famous *History of the Royal Society* published in 1667.[3] Sprat extols the then-new experimental science associated with the names of Galileo, Boyle and (later) Newton – the empirical drive coming out of the Scientific Revolution – explicitly in terms of its capacity to advance knowledge by long-term, progressive accumulation:

> [the members of the Royal Society] have endeavour'd to separate the knowledge of *Nature* from the colours of *Rhetorick*, the devices of *Fancy*,

or the delightful deceits of *Fables* … They have try'd, to put it into a condition of perpetual increasing … They have studied, to make it, not onely an Enterprise of one season, or of some lucky opportunity; but a business of time, a steddy, a lasting, a popular, an uninterrupted Work.

And he makes it clear that it is precisely by this steady advancement that the achievement of the new science will be of such immense practical use. He concludes his panegyric by contrasting this project with what the former, scholastic and Aristotelian approach could offer in this regard:

While the old could only bestow on us barren Terms and Notions, the New shall impart to us the uses of all the Creatures, and shal inrich us with all the Benefits of Fruitfulness and Plenty.

This prognostication has turned out to have been, on its own terms, very largely correct. Cumulative improvement in the material conditions of life driven by the growth of scientific knowledge and our associated technological capabilities has indeed been the main form taken by human development for the last three centuries. The steady business of 'perpetual increasing' in our knowledge of nature, which so impressed Sprat, is not founded merely in the preferences of a set of Restoration gentlemen weary of Aristotelianism and civil strife: it is inherent in the modern scientific enterprise. As Francis Fukuyama points out:

[I]f we look around at the entire range of human endeavour, the only one that is by common consent unequivocally cumulative and directional is modern natural science … The scientific understanding of nature is neither cyclical nor random … nor are the results of modern natural science subject to human caprice.[4]

It is true that Fukuyama oversimplifies this directionality for his own purposes. He claims in the same place that 'there are certain "facts" about nature that were hidden from the great Sir Isaac Newton that are accessible to any undergraduate physics student today simply because he or she was born later', without noting the important role of scientific paradigm-shifts in this process. Following the work of the historian of science Thomas Kuhn on the nature of scientific revolutions,[5] we should be familiar with the idea that there are facts about nature – say, those revealed by quantum physics – which weren't available to Newton, not simply because later generations have found these things out by standing upon his shoulders, but because within his fundamental framework of understanding they logically couldn't have been *formulated* as facts – in just the same way that Newton knew facts which couldn't have *been* 'facts' for Aristotle. But the main point in relation to progress still stands. Paradigm-shifts are triggered when having to account for the appearances comes to put too much strain on the established paradigm (as, most famously, discoveries in astronomy did on Aristotelian and then in its turn on Newtonian physics), and this happens through the gradual piling up of problems and discrepancies in the ordinary work of 'normal science' within each paradigm in turn. So scientific development, though uneven and sometimes proceeding by unpredictable jumps, is still in essence cumulative.

This cumulative directionality of empirical knowledge has then translated itself into social and economic terms in two principal ways. The first of these – ironically enough, given Sprat's innocent enthusiasm – has been through providing ever-improving weaponry for military competition, 'a great force for the rationalisation of societies', as Fukuyama wryly notes. But the second, corresponding more directly to Sprat's 'Benefits of Fruitfulness and Plenty', has been 'through the progressive conquest of nature for the purposes of satisfying human desires, a project that we otherwise call economic development'.[6] Familiarly, this project has proceeded not just by cumulative improvements in the machinery of production, but also through the associated rational organisation of labour, both together facilitating a positive-feedback process of continuous improvements in efficiency, expansion of markets and economies of scale. The result, of course, has been the material and social world we know in the urbanised West, where per-capita income has grown more than tenfold since the mid-1700s – and the conditions of which are now being decisively globalised.

But this model of progress has never been merely a matter of the techniques of production and its consumable products. Material progress also very importantly gave a new impetus and new opportunities to another basic human drive, intertwined with but fundamentally distinct from that towards 'fruitful plenty'. This is what Fukuyama calls the drive towards *recognition*, and identifies with *thymos*, the 'passionate' third part of the soul, alongside desire and reason, in the tripartite account of human motivation originally proposed by Plato in the *Republic*.[7] It is the sense of self-worth and self-respect that motivates each of us as individuals. Transformed into a social force, it drives towards the establishment of each member of society as an equivalent centre of dignity and value, respected and recognised as such by his or her fellows.

Fukuyama himself endorses a quasi-Hegelian account of the working-out of this drive for recognition through history, according to which it proceeds in parallel to the process of economic development, but by a different logic.

We don't need to follow him here, however, in order to appreciate the vital importance of the developing liberal-democratic order for the project of material progress, and vice versa. The ever-wider extension of literacy and education, for instance, is required for obvious reasons by an increasingly rationalised and technological society. Similarly, growth of individual responsibility and self-reliance is necessary to entrepreneurial initiative, and it is such initiative which can take advantage of the continually changing horizon of production possibilities subtended by technological development. The association of this growth of responsibility with the spirit of Protestantism, which also played such a vital role in initiating the Scientific Revolution, has been a commonplace of historical analysis since Weber and Tawney wrote about it at the beginning of the twentieth century.[8]

Corresponding to this extension of education and of individual initiative will inevitably be a growing sense of the *ethical* status of the autonomous rational individual, and of the need for society to be underpinned by certain ground rules for acknowledging and empowering its citizens as such individuals. Hence the impetus of economic development is so intimately bound up with the emergence of the liberal-democratic state, the modern domain of equality and rights, that we can't really conceive of human progress now without emphasising both dimensions.

Problems with progress

It is largely because our notion of progress and its value has this inextricably dual aspect, both economic and ethical, that the problems that have latterly emerged with the process itself and its consequences have proved so intractable. These difficulties only began to be recognised as critical, thanks largely to the emerging green movement, in the last quarter of the last century, although many of the deep human strains and tensions had already been identified by the far-sighted as early as the time of Dickens, Carlyle, de Tocqueville and Ruskin.[9]

The material aspect of the problem arises from the fact that economic development has classically been both conceptualised and institutionalised as a process of infinite expansion. In this, it responds directly to the underlying idea of progress identified by Nisbet: mankind has advanced, is advancing and *will continue to advance* – a structure of expectation that simply iterates itself open-endedly onwards. Correspondingly, conventional economic thought has long modelled the economy as a system unconstrained from without, where increases in production generate (along with higher consumption) greater household wealth which generates increased capital investment which generates further increases in production, and so onwards – the upshot being continually rising prosperity measured by ever-increasing Gross Domestic Product, or so-called economic growth. And despite the efforts of green economists from Georgescu-Roegen onwards,[10] that is how economic thought still very largely goes on modelling this system.

The conventional economic model, however, along with the government policies and business practices that it licenses, makes the kind of sense it makes only so long as two crucial assumptions go unchallenged. These are that both physical inputs to the production process in the form of natural resources, and the capacity of the natural environment to absorb and neutralise outputs in the form of wastes generated by production and consumption, are themselves open-ended. Those assumptions were as good as correct, for the time being, while the system was getting itself going in the eighteenth century, since the economy, even considered globally, was massively smaller in relation to the surrounding and enclosing environment than it was later to become. Even during dynamic capitalist expansion through the nineteenth and twentieth centuries, the assumptions still seemed to work in practice, though a few critics such as the thoughtful utilitarian John Stuart Mill could by then glimpse where things were heading.[11] But it has become increasingly evident over the last half century that the raw material sources and ecological sinks supplied by the natural environment (never, of course, *actually* infinite) have at last become an actively constraining factor. Physical limits that were always there in principle have begun to impinge in practice.

The signs of this impingement are those to which the green movement familiarly points. Rapidly expanding human populations, deploying the technological armoury developed by industrialisation in an increasingly determined pursuit of material well-being, have come to place huge and totally unprecedented demands on the natural resources of the planet. Not only has the abundance of raw materials been reduced in simple draw-down, but the hitherto self-maintained stability of ecological systems on which the whole intricate web of biospheric activity depends has been dislocated by pollution of all kinds, from agrochemicals to carbon dioxide emissions, and by the massive intrusion of megalopolis and mechanism into their workings. In pursuit of the ever-growing demands of urbanised economies, forests have been cut back, the oceans overfished, pasture-land laid under concrete and the absorptive capacities of the atmosphere and other environmental media grossly over-tasked.

The human population and its ecological support systems, that is, have got seriously out of kilter at the global scale, for the first time in history. Human needs for energy, food and raw materials have now so destabilised key life-support systems in the watercourses, oceans and atmosphere of the planet that their capacity to go on meeting current, never mind growing, levels of demand has been jeopardised, and the future material progress of the kind of global civilisation which we have built has consequently been imperilled.

These would be serious problems enough, were we able simply to abandon, in face of them, the mistaken assumption that indefinite economic growth is feasible within a finite physical system, and find a more functional alternative. Proposals for moving away from the infinite growth assumption and towards the steady-state model, originally canvassed (with remarkable prescience) by Mill in the nineteenth century, have been worked out in detail by a cohort of creative thinkers of whom the doyen has been the former World Bank economist Herman Daly.[12] It is true that any move to such a steady-state model, in which the premium would be on minimising rather than maximising material flows, would involve major social and institutional changes, requiring to be coordinated worldwide. It would therefore – as well as demanding unprecedented international political will and cooperation – take significant time, during which the accumulated ill effects of half-a-century and more of hyper-growth would still be coursing through the world system. But that, however daunting, is not the fundamental problem. The crucial difficulty with economic hyper-growth is that we seem to be locked into it, planet-wide.

In part this reflects merely the inner logic of traditional capitalism. Its characteristic incentive to increase the return on capital investment drives it to find ever-new ways to improve efficiency: this in turn is taken to mean constantly increasing labour productivity, so that less human input is needed for a given level of production, savings can be made on labour costs and prices can be kept competitive. But the human corollary of this process is its tendency to increase unemployment, which in turn leads to reduced consumer spending and possibly recession – *unless* permanent growth is always simultaneously taking up the slack with new opportunities for paid activity. Capitalism as traditionally conceived, in other words, has to keep running in order to stay on its feet.

Steady-state transformations of this pattern have been proposed. These typically retain the essential structure of the capitalist model while fine-tuning it. An excellent recent account is Tim Jackson's *Prosperity Without Growth*, which canvasses a number of options.[13] These include the maintenance of employment levels by reducing working hours without reducing remuneration, thus preserving effective demand; incentives to direct growth into sectors, such as some services, with comparatively low material throughput; and harnessing the drive for technical innovation to improving resource productivity, or 'ecological efficiency', rather than labour productivity.

But the difficulty that these proposed solutions to hyper-growth have to face runs much deeper than such regulatory and incentive proposals can reach. It reflects, in fact, those aspects of our attachment to material progress that are not merely a matter of classically recognisable economic behaviour. As Jackson himself puts it, in our world,

> stuff is not just stuff. Consumer artefacts play a role in our lives that goes way beyond their material functionality. Material processes and social needs are intimately linked together through commodities. Material things offer the ability to facilitate our participation in the life of society.[14]

This participation through consumption is supposed to offer us variously things to hope for, goals to aim at and dreams to dream, along with opportunities to construct would-be distinctive social identities in the attempt to placate those nagging questions about who we are and what it all means. And Jackson cites consumer, marketing and sociological research in support of the claim that material possessions are adopted into these identities not just for exhibition to others, but for our own self-understanding.

Correspondingly, this process of identification assimilates our rights to consume to what seem like much less disputable rights – to personal development, to freedom of expression, to democratic equality (since anyone can in principle acquire spending power), and so on. And by the same token it not only licenses us, but actually seems to require us in the name of justice, to extend those rights beyond the advanced economies to people in poorer nations still aspiring to enjoy the benefits of growth. (Recall in this context that there are now around seven billion people on the planet, the large majority of them falling into this latter category, and there will shortly be nine billion.)

But the restless consumption that is the dynamo of material progress is not just in these ways quasi-rational. It is also profoundly pathological – and not just because the global exercise of these consumption-dependent 'rights' now seriously threatens the ecological basis for human and other life. Such consumption is very far from being merely functional in relation to either material needs or needs for social status and participation, as Jackson wants to claim. It is pathological, in all these dimensions, because it is starkly addictive – and like all addictions, radically unsatisfying, strengthening its hold on the addict through the need to go ever further and further in pursuit of satisfactions that it cannot actually supply.

'Vacuous unease'

This point calls up a whole critique of progress which is of much longer standing than the explicitly ecological, going back as I have noted to earlier stages of industrial society and a distinguished cohort of nineteenth-century social commentators. It is a critique that very easily lends itself to parody and to righteous dismissal as 'elitist' – that is, it insists on relevant human standards in other areas than those of which competitive sport is representative. It nevertheless cannot be evaded.

Material progress has improved the human condition in many ways. People at large, at any rate in the advanced societies, are now better fed, clothed, housed and educated than they have ever been, and perhaps even kinder and more tolerant in their social relations, broadly speaking. Their capacity for atrocious cruelty under pressure seems undiminished, but for many the pressures tending to release that capacity are generally less than they were. People live, on the whole, longer, less physically dangerous and much more physically comfortable lives. But all this has come at an enormous spiritual price – though the word 'spiritual' is itself part of what is wrong. (It is in fact very significant that the long emphasis on material progress has made it so hard to find the right non-material term, in any kind of secular discourse, for that of which it has robbed us.) The literary and cultural critic F.R. Leavis nevertheless pointed very clearly to what is at issue here when he wrote of 'the despair, or vacuous unease, characteristic of the civilised world', and noted that this comes of 'profound human needs and capacities that the civilisation denies or thwarts'.[15]

Use a phrase such as 'despair or vacuous unease' in this kind of context, and defenders of progress will typically expatiate, as though they were answering you, on the vastly increased scope for *fun* which goes along with all the undoubted material gains just acknowledged, and on the evidence that ordinary people in huge numbers are taking and relishing their opportunities for it. (Significantly, though, they won't call it *fun*, they will call it a high standard of living, or even human well-being.) But it has to be insisted by anyone concerned for genuine human well-being that a very great deal of the fun (as Aldous Huxley long ago anticipated in *Brave New World*[16]) is grievously impoverished in human terms. It is also – on the theme of addiction – characteristically structured as a form of escape from its own impoverishment. Some of these escapisms are at least superficially harmless – but others, not insulated from the former in any clear-cut way, are much more insidiously destructive.

In the comparatively harmless category, perhaps, comes the ever-increasing obsession of the modern world with spectator sport, whether accessed directly or, more typically, through the medium of the television. Even here, though, it should be apparent that harmlessness is a matter of degree – consider the opportunity costs of the 2012 London Olympics, which set the British public purse back getting on for £9 *billion*.[17] (This is more than the entire annual revenue budget of the UK Department for International Development, a comparison that ought on a little mature reflection to be found utterly scandalous.) Continuous with mediated sport are the other forms of electronic entertainment, principally television, though increasingly also the internet, to which very many people sit passively captive for what are now significant fractions of their lives – nine years out of the average adult lifetime in the UK, apparently, if one believes the most recent research.[18] The vast majority of this material, as any intelligent person who will face facts must see, is feebly undemanding simply as a condition of having to be made accessible to a mass audience, and its mental effects are comparable to the physical effects of living on an exclusive diet of processed breakfast cereal – so already a variety of actual harms is in the offing, not least those of the addictive process whereby ease and speed of rubbish nutrition, physical or mental, accustoms the appetite to expect, and then to demand, rubbish.

As can be seen from the phenomenon of 'reality TV', however, there is no sharp line to be drawn between watching humanly empty fictions, fantasies and other time-fillers, and the kind of actively dehumanising voyeurism that characterises more destructive escapisms. Nor is there any sharp line between such voyeurism, in its turn, and the increasing pervasiveness of pornography, which reduces sexuality to a predatory form of entertainment and so numbs one of life's most vital centres. Again, the addictive quality in all this depends on the way in which the various substitute enjoyments are so much easier than real human relations, so that appetite comes to seek the substitute, and when it is found to be unsatisfying even as a substitute, to crave progressively more and more of it. Related to this, of course, and going on *pari passu* with the seepage of addictive expectations into ordinary acceptance, is the wholesale sexualisation of advertising and other forms of popular culture, and the associated (supposedly liberated, but actually deeply consumerist) approach to human sexuality as if it offered a menu of equally eligible 'lifestyle choices'.

That these conditions of life represent impoverishment by comparison with any past cultural phase should not really need arguing – at least, not to anyone genuinely acquainted with what life was and felt like in past phases, although this becomes an increasingly difficult point to make as real engagement with the significant literature and art of the past, where such acquaintance can be pursued, has become less and less common. A simple indicator that can be registered without any sophisticated cultural equipment is the evidently degenerated condition of ordinary sociability in the advanced societies – the now standard assumption that this must involve not just routinely excessive alcohol consumption, but also crude and insistent rhythmic noise. The inescapability of the latter in public and many private spaces offers a clear marker, to anyone with ears, of the increasingly widespread need to escape from our own life-impoverishment.

Insistent escapism, in fact, is the keynote that links all these forms of restlessly vacuous uneasiness to the actual physical restlessness that is also now such a pervasive feature of the societies spearheading 'material progress'. Obvious phenomena here are the need for constant individual movement reflected in wholesale dependence on the private car, and relatedly the 'tourist' explosion – the vast numbers of people churning between countries and across continents, partly in mere emulation of their neighbours and partly in the forlorn hope that life can be given point, at least for a little while, if we set ourselves to go somewhere novel and to stare at something outside our normal range of experience. (Genuine exploratory participation in the life of a foreign culture, requiring serious time-commitment and often self-sacrifice, is of course a different matter.) Here the dynamic coupling between pervasive escapism and the material destructiveness of progress – its ramifying ecological consequences – becomes unignorable. All this oil-fuelled, carbon-intensive mechanical motion, much of it now standardly airborne, plays a very large part in climate and ecological destabilisation. This destabilisation is thus seen to be engendered not just by the arguable benefits and psychological functionality of ever-increasing consumption, but also – and at least as vigorously – by the downsides which for want (as yet) of a better word we must continue to call spiritual.

Common characteristics of all these downsides are absorption of attention, provision of a temporary goal (its temporary nature concealed, temporarily, by the absorption), and addictiveness imposing an *ersatz* shape and direction on life, which then requires to be acted out in some form of further consumption. But always, and at the same time, all this activity is tacitly recognised as profoundly unsatisfactory, as the denial and thwarting of fundamental human needs for creativity, real purpose, loving intimacy, rootedness, imaginative and emotional richness and subtlety – failure of which to be met in a world of constant material improvement is what the activity is trying to compensate for in the first place. Leavis's 'vacuous unease', the spiritual hallmark of a civilisation which has 'progressed' materially beyond anything imaginable in the human past, is the irresistibility of the unsatisfying permanently compelled by the unsatisfactoriness of the irresistible.

Progress ecologised

The green movement, as I have already noted, has been cogently and force-fully identifying the various forms of ecological damage associated with material progress for over fifty years, ever since Rachel Carson's classic *Silent Spring*.[19] Alerting people to this damage, to its potentially devastating consequences for the biosphere worldwide, and to the urgent change of course that it made necessary, was one very important purpose for which this movement arose.

But it was not the only purpose. Another classic document of the movement, the 1972 *Blueprint for Survival*, is very revealing in this respect. Its authors state that:

> The principal conditions of a stable society – one that to all intents and purposes can be sustained indefinitely while giving optimum sat-isfaction to its members – are: (1) minimum disruption of ecological processes; (2) maximum conservation of materials and energy – or an economy of stock rather than flow; (3) a population in which recruit-ment equals loss; and (4) a social system in which the individual can enjoy, rather than feel restricted by, the first three conditions.[20]

We have here, plainly enough, an early version of 'sustainable develop-ment'; and this way of putting it makes the final condition seem clearly instrumental to the achievement of the first three. But actually the *Blueprint* devotes almost as much attention to what it calls social stability as to eco-logical. A whole Appendix is given over to a critique of contemporary civilised life. 'The vast and chaotic human societies in which we are living are by no means normal', it is argued. This abnormality is established by an anthropological, and so implicitly an ecological, criterion – modern society is disintegrated and alienated in ways that no previous human civilisation has been – so the dangers that it poses to our survival are clearly linked to those stemming from our biospheric situatedness. Abnormality is explicated, however, precisely in the terms of vacuous unease that we have just been rehearsing. These chaotic societies are 'characterised by a general feeling of aimlessness, a frantic, almost pathetic search for originality, over-preoccu-pation with anything capable of providing short-term entertainment, and beneath it all a feeling of hopelessness and of the futility of all effort'.[21] Decentralisation, the rebuilding of local community and associated recon-nection with the natural environment are put forward by the *Blueprint*, if one reads it as a whole, at least as much in answer to these problems as they are intended to help meet the demands of the three 'ecological' conditions.

The *Blueprint* represented a pivotal moment for environmentalism in more ways than one. We know what subsequently happened to the sustainable development element of the analysis: it quite rapidly overcame initial resistance and was welcomed into the mainstream. Offering in its terms of reference something readily recognisable to scientists, policymakers and eventually the general public, it was in important ways *unthreatening* despite the warnings of future damage with which it was loaded. It was especially unthreatening because it made possible the posture already analysed in Chapter 1 – serious apparent concern for change to avoid future damage, combined with a tacit commitment to not changing too drastically for present comfort.

What happened to the analysis of vacuous unease? Parts of the movement have indeed continued to press the longer-standing cultural critique of progress in parallel to the ecological critique. But, significantly, this has been articulated much less confidently and unanimously than the case against ecological damage. The voices strongly emphasising this critique within the green chorus have been comparatively few: one thinks of Schumacher, Satish Kumar and Theodore Roszak, and more recently of Bill McKibben in his earlier writings,[22] but other names of equal stature do not come readily to mind. Nor are the reasons for this collective hesitancy obscure. Ecological damage is a practical consequence, in principle detachable and remediable, of material progress – but vacuous unease is its spiritual signature. You can disavow the former, but hardly the latter, while still remaining attached to, and driven by, impulsions deep in the progressive mindset. And the standard aspirations of the 'green turn' as it has been mainstreamed represent progress *ecologised*.

That is, they look to the continuation of progress by other means. Perhaps ever-increasing comfort and abundance, at the kinds of level to which the West has become accustomed, are no longer going to be possible. But if we could pull off the sustainability trick, the magic combination of renewables and smart technology and green jobs and global cooperation – if we could accommodate to the limitations that our natural situation and ecological resources place on us, instead of trying to ignore them – then we should still be able to go on engaging in the familiar pursuit of both comfort and justice (materially conceived) as sufficient human goals. We could continue to suppose that a progressively more equitable distribution of wherewithal to meet the material needs of life, and of opportunities such as education closely related to meeting those needs – with, beyond that, the worldwide elimination of poverty, hunger, violence and avoidable ill health as longer-term prospects – made sense of human existence.

But is not this kind of aspiration wholly admirable? Is not progress made sustainable an inspiring ideal, a noble vision of human possibilities?

Actually, the aspiration is corrupt. We have already seen how sustainable development as a political paradigm is deeply implicated in bad faith through its characteristically two-faced affordances to the culture of denial and willed optimism. But that in turn reflects a more fundamental failure: a corruption at the level of being, deep-seated in the possibilities of consciousness under the present conditions of human life.

Much of the rest of the book, in particular Chapters 3 to 6, is about explicating and defending this claim. That will require a good deal of conceptual work. At this stage I must confine myself to indicators that might helpfully scope out the task. Luckily, as regards the mainstream version of sustainability and the progressive aspiration, there is a powerful one ready to hand – a fascinating icon of complex corruption – in Ian McEwan's recent novel *Solar*.

Here is McEwan's protagonist Michael Beard, a rather implausible Nobel Laureate in physics, looking out of the window of his plane as it circles in a stack above Heathrow:

> The fields and hedgerows, once tended by medieval peasants or eighteenth-century labourers, still visibly patterned the land in irregular quadrilaterals, and every brook, fence and pigsty, virtually every tree, was known and probably named in the Domesday book … And ever since, named again with greater refinement, owned, used, costed, traded, mortgaged; mature like a thick-crusted Stilton, as richly stuffed with varied humanity as Babel, as historical as the Nile Delta, teeming like a charnel-house with ghosts, in public discourse as dissonant as a rookery in full throat. One day this brash and ancient kingdom might yield to the force of multiple cravings, to the dreamy temptations of a giant metropolis, a Mexico City, Sao Paulo and Los Angeles combined, to effloresce from London to the Medway to Southampton to Oxford, back to London, a modern form of quadrilateral, burying all previous hedges and trees. Who knew, perhaps it would be a triumph of racial harmony and brilliant buildings, a world city, the most admired world city in the world.
>
> How, wondered Beard, … could we ever begin to restrain ourselves. We appeared, at this height, like a spreading lichen, a ravaging bloom of algae, a mould enveloping a soft fruit – we were such a wild success.[23]

Once landed, Beard goes on to address a conference of pension fund managers on solar power, in which he has a financial as well as a scientific interest:

We either slow down, and then stop [burning fossil fuels], or face an economic and human catastrophe on a grand scale within our grand-children's lifetime… How do we slow down and stop while sustaining our civilisation and continuing to bring millions out of poverty? … the answer to that burning question is … affordable clean energy.

There, very vividly in those two linked passages, we have 'progress ecol-ogised': the impetus onwards and upwards, the powerful thrust towards achieving 'racial harmony and brilliant buildings' while 'continuing to bring millions out of poverty' – that is, still towards continuing material better-ment and its associated forms of justice – all driven by what is really pushing megalopolis irresistibly across the landscape, the 'dreamy temptations' of our longstanding inability to restrain ourselves from all our variegated forms of developmental busy-ness.

Given present conditions, as recognisable by a scientist (Beard is at least convincingly that), the obvious imperative is then to shift our unrestraint decisively onto a ('sustainable') basis of renewable energy. But, as the whole novel also makes sufficiently plain, what really drives this unrestraint, this algal efflorescence of 'development', is the attempt to compensate for life-emptiness, of a kind that its protagonist himself dramatically exemplifies – he is addicted to junk food, increasingly (and by the end grossly) over-weight, a feckless womaniser, a ruthless plagiariser in his intellectual work: a complete, all-round, symbolic human mess. But he is, in all this craving unease, only an exaggeration for effect of ways of living now prevalent, as we have seen, in the so-called 'advanced' societies: the comfort-materialism and the escape-indulgence – and, what the novel clearly brings out as an inevitable corollary, the lack of any real compass for action except self.

Beard, in classic bad faith, both knows that he is trying to escape himself and doesn't know it, avoiding the recognition in ever-further half-recog-nised avoidance; and meanwhile, his self-absorption ineluctably brings on the disaster that ultimately sabotages his almost-breakthrough in planet-saving photovoltaics. Veering uncertainly between comedy and something else, the novel contrives to make this outcome depend on his having opportunistically framed a love-rival for a murder which was in fact an accidental death – but creaking literary machinery is beside the point here. What is vividly clear is the way unease, escapism and ecologised progress are intimately linked in a nexus of deep corruption.

But why, one is driven to ask, would human beings ever endure to live like this, pursuing lives of unsatisfying and lonely self-destructiveness of the kind that Beard symbolically embodies, and in the process threatening to create such a cancerous urban wen as that envisaged 'world-city' sprawling across half the Home Counties, even if it *could* be rendered sustainable? Why, unless they were attempting to escape from some *radical* terror? *Solar* shirks this issue – McEwan, whether or not deliberately, leaves the question hanging over the novel as it peters out, and it will hang over my own argument until we are in a position properly to address it.

Beyond consumerism?

Beginning to do so will be the task of the next chapter. Before moving on to it, however, I want to reinforce my claim that environmentalism standardly aspires to 'progress ecologised'. For after all, Beard's preferred techno-fix represents only one kind of proposed solution to our difficulties. His dream of solar power on a grand scale, enabling present patterns of civilisation not only to continue but to go on extending themselves, can stand in for the wide variety of such techno-heroic aspirations now being canvassed with increasingly desperate urgency, as we have already seen from Lynas's *The God Species*. But what about that strand of green advocacy that talks about downsizing, the virtues of slowness and localism replacing the megalopolitan buzz, the whole idea of what Tim Jackson calls 'flourishing within limits'? Jackson links such flourishing to the philosopher Kate Soper's notion of an 'alternative hedonism'; as he expresses this:

> Anxious to escape the work and spend cycle, we are suffering from a 'fatigue with the clutter and waste of modern life' and yearn for certain forms of human interaction that have been eroded ... A shift towards alternative hedonism would lead to a more ecologically sustainable life that is also more satisfying and would leave us happier.[24]

This vision, surely, cannot be plausibly characterised as the continuation of material progress by other means? It seems to be premised on a deliberate turning away from the values that drive this kind of progress – competitive acquisitiveness, the measuring of one's success by the size of one's mortgage or car, the desire, in Erich Fromm's distinction (made a good while ago but still very relevant), to *have* rather than to *be*.[25] This kind of consciousness, celebrating what are now typically designated 'intrinsic' values such as self-acceptance and communal belonging, points us towards the good life in the steady-state economy – 'prosperity without growth', or at least without the continual growth in material throughput (from raw materials extracted through to wastes needing absorption) on which conventional progress depends and which has so damaged the biosphere. This, surely, is the *real* green answer to the ills of progress?

Well, up to a point … The problem is that all this is claimed to represent a move *beyond* consumerism, when actually it is no such thing.

The post-consumerist claim is made quite explicitly by Soper:

> A changed perception of the 'good life' (and 'alternative hedonism') is … an essential stimulant to any possible future curbing of consumer culture … a post-consumerist vision … in which enjoyment and personal fulfilment are indissolubly linked to methods of production and modes of consumption that are socially just and environmentally protective.[26]

This comes in a book focussed specifically on rethinking consumption in its relation to citizenship. The point has always been implicit, however, within this tradition of green advocacy. Jonathon Porritt has long been a significant voice here, both in his own writings and in the work that he has driven through Forum for the Future.[27] Right back in the 1980s (it is startling to reflect how short a while ago that actually is, given the green movement's subsequent rise to prominence), Porritt's *Seeing Green* was endorsing voluntary simplicity as 'a way of life … with the emphasis more on human relationships and personal development than on material consumption'. Latterly, he has assembled in *Capitalism as if the World Mattered* (2005) a great deal of more recent thinking along the same lines. A vision document from the Forum that seeks to set out what a sustainable society might look like proposes that in such a society 'the ambition of politicians and community leaders is to ensure the highest possible quality of life within the operating limits of the natural world'.[28] This, argues Porritt, is not the mere day-dream which it might appear in one's more cynical moments, because it now answers to a substantial and growing pressure towards 'inner-directedness' in the West. He quotes a definition of this category as

> people whose prime motivation is no longer conspicuous consumption or keeping up with their neighbours, but autonomy, self-expression, health and independence. These are people who are suspicious of mass-production, who want things customised or tailor-made … who are definitely part of the world of self-actualisation.[29]

And still more aspirationally, drawing on the work of Herman Daly already noted, he presents an expanded 'five-capitals' framework for sustainability capitalism (fully recognising natural, human and social as well as manufactured and financial capital), in which the avowed ends served by the whole system are not – or, are only mediately – consumer goods and services; ultimately, they are 'happiness, harmony, identity, fulfilment, self-respect, well-being, transcendence, enlightenment'.

Put like that, how could one cavil? Indeed, contrasted with the havoc that material progress is currently wreaking, this is an uplifting vision. Porritt's own life-work of bringing the political and economic arguments for it in from the fringes and making them seriously respectable has been heroic. But for all that, the deep problems with tailor-made self-actualisation should not be hard to see: and they are problems that, properly considered, leave one much less sanguine that the dilemmas of progress are really being left behind on this trajectory.

Of that portfolio of non-material ends, we may take *fulfilment* as a suitably workaday summation. It seems to presuppose identity and self-respect, gives a definite human shape to happiness and well-being, and represents whatever of solid value there may be in harmony, transcendence and enlightenment. The core difficulty is then that fulfilment *pursued as an end of life* is necessarily unachievable. Of course, we all know people who seem to be fulfilled. But that is precisely because they don't see the purpose of whatever they are doing as 'enjoyment and personal fulfilment', but as the discharging some obligation or getting done some job that is big enough to fill a human life. As soon as fulfilment itself (what Soper sometimes calls, in an awkwardly revealing phrase, 'self-pleasuring') becomes the focus, its capacity to serve as a plausible human end simply drains away. And this remains true however non-materialist the satisfactions from which it is supposed to be derived. Indeed, *trying* to fulfil oneself through 'human relations and personal development' – that is, using others (and oneself) essentially as *means* to fulfilment – is likely to lead to more grievous disappointments, and often as a consequence to more insidious and vicious forms of compensatory damage, than even the forlorn attempt to find a point to life in shopping.

What follows is that the alleged post-consumerism of pursuing 'the highest possible quality of life within the operating limits of the natural world' remains structurally consumerist. It is worth pressing a little on the underlying analogy here in order to enforce this point. Ordinarily, and naturally, we *consume* in order to fill ourselves up, and then we stop until equally natural processes of digestion and evacuation have done their work and we need to consume some more. Unintermitted consumption, the constant restless stuffing ourselves with forms of gratification supposedly to be derived from products and services (the dynamic on which late capitalism depends), can therefore only proceed, one might suppose, if there is an equally constant emptying out going on at some metaphorical other

end. That emptying or draining away of significance, the permanent seepage of satisfactoriness from the permanently irresistible that drives 'vacuous
unease', must proceed through what might be called (slightly flexing the
metaphor for delicacy's sake) some kind of hole in the human soul – some
gap at the core of human being through which our attempts to identify
activities or possessions or achievements as giving meaning to our lives are
always vanishing. But if that is so, any kind of life-significance pursued in
this maximising spirit (the 'highest possible' quality of life) is going to be
in the same case. It is no matter whether the achievements that we have
in view here are quantitative or qualitative: even the significance of harmony, transcendence and enlightenment, if proposed as possible objectives
of maximising pursuit intended to give us fulfilment, will drain constantly
away – and the craving for them will be as constantly renewed in a dynamic
of permanently unfulfilling consumption.

All this, of course, is elementary life-knowledge for which the culture
used to have a speakable language, but which has become strange-sounding
and much harder to articulate as that language has been ('progressively')
eroded. I attempt more of the work of articulation in the next and subsequent chapters. The point in the immediate context concerns 'the operating
limits of the natural world'. Substituting the maximising pursuit of fulfilment for that of ever more material consumption, 'prosperity without
growth' for the growth economy, is supposed to do what would enable us
to keep within those constraints. Actually, what it is much more likely to
do, given the unescaped 'qualitative' consumerism that it still represents, is
to provide us with yet greater opportunities for bad faith in our dealings
with limits. I have quoted elsewhere Mark Lynas's justifiably jaundiced
view of lifestyles involving 'driving to the supermarket in a hybrid 4 × 4
to buy organic carrots, somehow believing that we are part of the solution and not part of the problem'.[30] But more than merely this order of
mundane evasion is at stake. What we should clearly *not* have done, in the
steady-state economy as thus far conceived, is to have placated the restless urgency that drives contemporary forms of progress. We should not
have loosened the underlying conviction that ever-closer approximation to
some desired goal, so that tomorrow always has the potential to be on some
dimension *better than today*, is what life has to be about. Remaining in the
grip of that conviction, we shall not be secure against the unsatisfactoriness
built into consumerism (permanent pursuit of 'fulfilment' of *any* kind) – so
that, as we find the 'highest possible quality of life' still failing to fulfil us
in the advanced societies, we shall be driven to pursue ever-greater and
more hubristic technological refinement in the means of its delivery, and
to insist ever more emphatically on its extension to the burgeoning billions

in the rest of the world (putting the techniques of eco-efficiency under still greater demands). We are not sufficiently far here, even on the would-be alternative 'post-consumerist' track, from Beard's techno-heroic vision of racial harmony and brilliant buildings. This whole prospect, in fact, offers us a scenario neither of managed ecological modernisation nor of steady-state equilibrium, but of inherent and self-feeding instability. As such, its chances of respecting the limits of the natural world do not look good.

★ ★ ★ ★ ★

We began this chapter by tracing the evolution of 'progress' in the hope of identifying where eco-activist denial was really coming from, and what was reinforcing it so strongly. The profile of pressures and tensions here should now be apparent. Green solutions as standardly propounded represent the continuation of material progress by other means, because progressivism remains a deep-seated human craving even when we have seen what its hitherto favoured forms have done, and are doing, to the biosphere, the climate and the prospects for life on Earth. So, if the only way to give ourselves the chance of continuing progress is to find sustainable green solutions that will also begin to decouple this process from damage, there *simply must* still be time to find them. But, in turn, the only way to preserve such a belief against the encroachments of scientifically literate realism, on which eco-activists themselves have long relied, is denial — whether this takes the form of dogged persistence or of technological hyper-optimism. And hence — since that denial is also at the same time self-recognised as incredible — the identity-crisis of environmentalism which we started by noting follows on.

But in that case, it seems that we have only pushed the problem of explaining this crisis further back. If a deep attachment to progress is driving denial, what in turn fuels and fosters that deep attachment — particularly in a movement premised, at one level, on recognising the destructiveness of progress? If we are really going to understand what is happening, that is the question to which we must now turn.

Notes

1 See Nisbet (1980), pp.4–5.
2 See Lasch (1991).
3 Sprat (1667/1959); the two quotations are from pp. 62 and 438 respectively.
4 Fukuyama (1992), p. 72.
5 See Kuhn (1962).
6 Fukuyama (1992), p. 73.
7 See Fukuyama (1992), Chapter 15 onwards.
8 See Weber (1905/1992), Tawney (1926).
9 The classic texts here, of which there are many editions, are Charles Dickens, *Hard Times* (1854); Thomas Carlyle, *Past and Present* (1843); Alexis de Tocqueville, *Democracy in America* (1835, 1840); John Ruskin, *Unto This Last* (1860).
10 See in particular Georgescu-Roegen (1971).
11 See Mill (1848/1994), Book IV, Chapter VI, 'Of the stationary state'.
12 The key text is Daly (1992).
13 Jackson (2009).
14 Jackson (2009), p. 98.
15 Leavis (1975), p. 10.
16 See Huxley (1932).
17 For the sum quoted, see *The Guardian*, 23 October 2012, available at www.theguardian.com/sport/2012/oct/23/london-2012-olympics-cost-total. For the DFID annual budget, see the HM Treasury document *Spending Round 2013*, available at www.gov.uk/government/uploads/system/uploads/attachment_data/file/209036/spending-round-2013-complete.pdf
18 See for instance www.dailymail.co.uk/news/article-2235007/Average-Briton-spends-years-life-watching-TV.htm
19 Carson (1963).
20 From Goldsmith *et al.* (1972), p. 30.
21 Goldsmith *et al.* (1972), p. 111.
22 See in particular Schumacher (1974); Kumar (2008); Roszak (1981); McKibben (2010).
23 McEwan (2010), p. 110. The second quotation, immediately following, is from p. 149.
24 See Jackson (2009), p. 148.
25 See Fromm (1978).
26 Soper and Trentmann (2008), p. 199.
27 See Porritt (1984) – the quotation is from p. 204 – and Porritt (2006).
28 Porritt (2006), p. 295.
29 Porritt (2006), p. 298. For the five-capitals framework, see also pp. 112–16.
30 See his article 'Can shopping save the planet?', *The Guardian*, 17 September 2007.

3

PROGRESSIVISM

What had started to emerge by the close of the last chapter was the *mythic* status of progress in contemporary consciousness.

The phenomena of entrenched eco-activist denial which we have been exploring make it apparent that belief in material progress as a permanent human project isn't just an ordinary illusion, something to be abandoned once it has been shown to be ungrounded. If that were the case, the glaring implausibility of the belief that we are still in time to save the planet from global warming would have forced us to question the belief of which that belief is the defence – that something like 'sustainable development' can indeed reconcile material progress with ecological constraints. Exposure of that belief as illusory would then have forced us to question the belief that it was meant to transpose into contemporary conditions of ecological awareness – that is, the underlying belief that material progress can be a permanent and indefinite human good. As each of these beliefs had been revealed in turn to be unstable, so the broader belief that it was trying to rescue would have been exposed. The long commitment to progress – based, originally, on the then-plausible belief that developing technologies could be applied indefinitely to effectively inexhaustible resources – would have had to be recognised as the long illusion, beyond the ending of which we now had to think. And the question 'What comes after this illusion?' – that is, after sustainability – while hard enough to answer, would not have been as hard as it actually is to *ask*.

But myth does not behave like this. Myth and illusion, while they occupy overlapping epistemic territory, are not the same thing. Standardly, an illusion is a false belief mistaken for fact; once that mistake has been identified, it has no further claim to credit. (We talk, typically, of its having been dispelled.) A myth, on the other hand, is an exemplary narrative that is felt to embody certain deep truths, and that is thus capable of retaining its hold on us even after its exposure as insufficiently grounded. A belief, or a narrative nexus of beliefs, which has attained to mythic status is tenacious in the face of evidence that would dispel an ordinary illusion, precisely because it claims the allegiance of our understanding by expressing for us commitments going deeper than the reach of ordinary evidence. Myths may over time crumble and disintegrate – accompanied by much psychological trauma as they do so – but they are not *dispelled*. Often, indeed, such disintegration happens only because they are being replaced by other, alternative myths.

In the grip of a myth

Recognition that belief in progress is in this way mythical has been rare in contemporary discussion, but not unprecedented. Its most forceful recent advocate has been the political theorist John Gray, who describes it compellingly in his book *Heresies* as one of those 'beliefs to which we cling for reasons that have nothing to do with the truth'.[1] The paradigm of that kind of belief, for any rational secular modern, is very evidently religion, and Gray has no hesitation in making the comparison. Moreover, it is one that for him is in many ways to religion's advantage: 'Whatever role it may have had in the past, belief in progress has now become a mechanism of self-deception … In contrast, the myths of religion are ciphers containing the truth of the human condition.' He pushes this comparison and contrast still further with the argument that resolutely optimistic progressivism is self-deceptive because it is a kind of unacknowledged repression – and what is being repressed is, precisely, the understanding that it is meeting a quasi-religious need:

> Secular societies are ruled by repressed religion. Screened off from conscious awareness, the religious impulse has mutated, returning as a fantasy of salvation through … science and technology. The grandiose political projects of the twentieth century may have ended in tragedy or farce, but most cling to the hope that science can succeed where politics has failed: humanity can build a world better than any that has existed in the past. They believe this not from real conviction but from fear of the void that looms up if the hope of a better future is given up. Belief in progress is the Prozac of the thinking classes.

Both in *Heresies* and in his earlier *Straw Dogs*, Gray develops a powerful critique of faith in progress based on this identification of its mythic, quasi-religious status. Progress itself may bring many benefits. His favoured example of anaesthetic dentistry is sufficiently telling: we have managed to improve dramatically on the situation, noted by de Quincey in the early nineteenth century, where a quarter of human suffering was toothache.[2] There are plenty of similar examples of material progress relieving pain or enhancing pleasure, to be set against the equally undoubted psychological and ecological damage that it has also been doing. But it is the abandonment of precisely this dispassionate comparison of benefits and costs, in favour of *faith* in progress, as offering meaning and hope, that the mythic takes over. According to Gray, faith in the cumulative growth of scientific knowledge, as now our only substitute for Providence, is what leads us into all the contemporary follies of *homo rapiens* and towards the next, anthropogenic, mass extinction.

This argument is very congenial to my own approach, and Gray makes it with an aphoristic vigour which I quote with gratitude. But for all its force and cogency, it remains essentially a striking comparison enforced and illuminated, rather than an explanation. And explanation is what we are surely owed here. For of course, faith in material progress, and the role played by such faith in our lives, are also very *unlike* how things were with any historical religion. However striking the analogies, faith that science can build us a better world differs markedly from any kind of faith in a numinous Providence, and in at least one crucial respect, startlingly so: it can be seen to be pretty *evidently* misplaced, if we will only look at the empirical reality of our condition. (This is, or should be, the standard response to all those people who talk glibly about 'the success of science' – point them gently but firmly back towards the state of the world around them and ask: *what* success, exactly? The criteria which this evokes will rarely stand examination.) Religions, moreover, as Gray acknowledges, spoke to a deeper set of human worries than concern for material progress seems, at least on the surface, able to touch – and this is true even of those varieties of religion that were most conducive to individualistic self-reliance and self-improvement. Prozac, whether metaphorical or literal, is so palpably *ersatz* a substitute even for Protestantism that we are at risk of failing to understand how the substitution could possibly have taken place, unless we can identify some human concern sufficiently radical for both these very different kinds of manifestation to be traceable back to it.

Tim Jackson (as we have seen, a committed defender of progress by ecologising it) nevertheless points us in the right direction here when he writes of prosperity, albeit decoupled from growth, as a vision of human progress. Prosperity speaks of the elimination of hunger and homelessness, an end to poverty and injustice, hopes for a secure and peaceful world. And this vision is important not just for altruistic reasons but often too as reassurance that our own lives are meaningful. It brings with it a comforting sense that things are getting better on the whole – rather than worse – if not always for us then at least for those who come after us. A better society for our children. A fairer world. A place where those less fortunate will one day thrive. If I cannot believe this prospect is possible, then what can I believe? What sense can I make of my own life?[3]

Gray himself gestures in the direction of 'reassurance that our lives are meaningful' when he writes of 'the void that looms up' if aspirations to progress are abandoned, and also when he claims in a later passage that 'believers in progress are seeking from technology what they once looked for … in religion: salvation from themselves'.[4] But even these remain only gestures, and we look in vain for anything in his critique of progress that explores what they might be gesturing towards.

Yet that is surely not so far to seek, though the actual exploring of it will take us into philosophical, indeed metaphysical, rather than political–theoretical terrain. I want to argue that the void from which we are trying to escape and recoil from, which gives the myth of progress its compensatory grip, is our modern inability to make sense, as individual conscious agents, of the fact of personal death. This is what turns the plausible thought that a better future matters in some important ways, into progressiv*ism*, the essentially mythic conviction that *only* a better future can give point to life. That inability to confront our mortality reflects the collapse of institutionally religious aids and explanations that could once be taken seriously, and the patent inadequacy of the various individualistic spiritualities on offer as substitutes. In turn, it explains why we are so hooked on material progress that, confronted at last by its full environmental consequences, we can only try to ecologise it; and also why the attempt to do so depends so heavily on entrenched refusal to see what is obvious. The deep denial in which the destructive drives of Western civilisation and counter-aspirations to its 'sustainability' are both complicit, is denial of the knowledge (which, somewhere, we all tacitly have) that the pursuit of a future of racial harmony and brilliant buildings, or however else we envision progress, comes nowhere near being enough to silence the deep questions ('What, ultimately, *for*? What can I live by?'), or even, for long, to defer them. And yet, as things stand, that pursuit seems to be all we have for hope – and hope is necessary for life.

Why focus on the confrontation with death here, rather than on other, apparently less drastic ways in which the issue of life-meaning might come up? The answer is that this is the level at which religions have traditionally acquired their firmest hold over us. The Buddhist scholar Stephen Batchelor observes that 'Religions are united not by belief in God but by belief in life after death':[5] that is, the essential impulse towards explicitly religious formulations stems from our profoundly disconcerting awareness, not permanently suppressible, that death is coming for each of us, that usually we cannot know when, and that most of our ordinary life-purposes are called in question by that recognition. Correspondingly, as we shall see, faith reposed in material progress offers a substitute for belief in life after death that goes with the grain of technologically driven materialism and its attendant conception of the human agent: a life-pattern tacitly premised on the assumption that the individual finitude from which we recoil can be transcended in the permanent open-endedness of indefinite human betterment.

Now the fact that human beings fear death ('as children feare to goe in the darke', added Bacon[6]) is hardly news. Nor is the fact that the decline of institutional religion has led us increasingly to repress rather than address that fear, so that in the advanced West death has replaced sex as the great unmentionable. And meanwhile, even given that lurking fear, many people seem well able simply to get on with their lives. We do not appear to ourselves to be driven *inevitably* into a cycle of addictive materialism and denial in a vain attempt to escape our mortality. But I believe that appearances are largely deceptive here – as of course they would be, should denial be playing anything like the role suggested. Whether or not this is indeed so, however, can only really be demonstrated by coming to closer grips with the issues.

'The anaesthetic from which none come round'

What is so bad about our own death, that we should each fear it as we do, and turn our faces so resolutely from it while we can?

A very longstanding response dismisses such fear as simply irrational. The ancient Greek philosopher Epicurus observed that we never actually encounter death, since 'where death is, I am not; where I am, death is not'.[7] In the twentieth century, Ludwig Wittgenstein re-expressed this as the point that death is not an *event in life*; my death is not something of which I as experiencing focus can be conscious, since necessarily while I am still conscious of anything it has not yet arrived.[8] If this evidently persistent idea is right, then it makes as little sense to fear my death as it would to limp because someone else's shoes pinch.[9] We might understandably fear *dying*, since this can be a physically painful process, and – if we are aware that it is happening – emotionally painful and distressing too, as we take leave of loved ones and they of us. But the one form of distress can often be palliated by medication and appropriate care, and we can at least strengthen ourselves by anticipation against the other. And beyond that, how can there be anything fearful about *being dead* itself, since we shall not be around to experience it?

This is a classic instance of an apparently decisive demonstration which leaves virtually no one convinced. Trying to explain why not, the contemporary American philosopher Thomas Nagel (among a good many others) has argued that, although it may be irrational to fear *being dead*, we are nevertheless perfectly rational in fearing to be deprived by death, whenever it occurs, of possibilities that would otherwise have remained open to us.[10]

But this in turn is surely true only up to a point. If, still in my sixties, I were to learn that I had only a few months to live, I should certainly then feel much grief at missing the opportunity to see and get to know all those grandchildren who haven't come along yet; and for fear of that grief I might now fear finding out that I had a terminal illness, quite apart from fearing its potential painfulness. In my late eighties, however, should I have the dubious fortune to reach them, it seems quite plausible that whatever possibilities then remained genuinely open to me might not offer sufficient compensation for the disability, restrictedness and discomfort of daily living. Once the game had really ceased to be worth the candle, it could seem perfectly reasonable to fear *not* developing a suitably terminal illness. (This is the thought that used to be expressed, before the questionable triumphs of modern geriatric medicine in helping people survive themselves, by the description of pneumonia as 'the old man's friend'.) Yet still in that contemplated situation, where my actual specific possibilities are not only not going to be regretted, but are such that I can suppose myself quite cheerfully forgoing them, death as the terminus of all the options remains something from which the mind instinctively recoils.

I think the poet Philip Larkin comes much nearer the truth of this kind of recoil than any of the philosophers whom I have quoted. His late piece 'Aubade', one of the starkest meditations on death ever written in its dry, utterly unheroic matter-of-factness, bleakly characterises our terror of extinction as recoil from the prospect of

> … no sight, no sound.
> No taste or touch or smell – nothing to think with,
> Nothing to love or link with,
> The anaesthetic from which none come round.[11]

That last line focuses the issue with laconic brilliance. Under a general anaesthetic, consciousness and agency are suspended – but I can make sense of this absence of myself, this gap of nothingness in my life, just because it *is* a gap and understood as such against a background of continuing presence. My consciousness is temporarily cancelled, but I will come round and normal service will be resumed. Death, however, is the last cancellation, the one after which there will be no resumption. And yet I cannot help conceiving even of this final absenting of myself as a permanent gap in *my consciousness*. That is both strictly unintelligible (there can't be an intermission going on for ever) and radically appalling – the permanent deprivation, not of just specific possibilities but of possibility itself, has to be apprehended within a continuing space of possibility. Anticipated death suddenly seems disconcertingly like a logical version of being buried alive.

But the fear of this prospect is, precisely, not a physical but a *metaphysical* fear: that is why accounts that attempt to dismiss or corroborate it in terms of ordinary experiential fears miss the point. (The way they do so also helps to clarify what the term 'metaphysical' means in this connection: metaphysics is *out of the ordinary*, but only because it is the kind of thought-exploration by which we try to trace the inner and outward conceptual boundaries of ordinary experience and make sense of where they lie.) In fact, what outcrops here is the radical division within our concept of personhood, between myself as first-personal subjectivity and myself as part of the ordinary third-personal physical world. This division also informs the most intractable philosophical problems, such as those of free will and of consciousness: How can there be genuine options for the subject in a deterministic universe? How can the fundamentally material generate a perspective on itself? Awareness of one's oncoming death, however, is where the crux which in most discussions of these long-familiar questions is merely theorised, has to be actually lived by each of us. And what that suggests is that, if our would-be compensatory attachment to material progress is indeed motivated by recoil from that awareness, it too will need an explanation of the same metaphysical order.

Larkin's insight into our fear of death is a poet's. In pursuing a metaphysical account of what is going on, however, we need a more discursive approach (which really means, one relying on more flat-footed metaphors). So what, in prose, will happen when I die? A certain organic entity will cease its vital functioning, and the processes of decay and disintegration to which it is (alas) already subject will accelerate rapidly. But also: this intimate inward subjectivity, the continuing first-person consciousness that I tend to think of as essentially *me*, will no longer exist. While some objects *in* the world will alter their form, that is, a unique perspective *on* the world – and one to which I am uniquely attached – will be utterly abolished.

This notion of a lost unique perspective is already odd. In a sense, of course, any perspective is unique: you can only see the mountain *just* this way from just *here*. But if we are thinking of a perspective as something like this, on the model of the 'viewpoints' that get marked on walkers' maps of the Alps, then it couldn't be lost or abolished just by virtue of the fact that no one any longer looked from it, even if that remained the case for ever afterwards. That point in space, and that particular angle on what could be seen, would still be available. (And if there were an earthquake, or the planet exploded, all that would affect would be whether the viewpoint could in practice be occupied and what the prospect from it would then be.) But the perspective which will cease to be available when I die is unique in the quite different sense that only I, as subject of my consciousness, can ever occupy the relevant standpoint. The perceptions, feelings and motivations that constitute it are logically private – no one else can have them, any more than anyone else can think my thoughts, feel my pains or sneeze my sneezes. Thus, when I have become permanently unable to occupy that standpoint, it looks as if it must vanish absolutely. The difference is that between 'There will no longer be anyone here' and 'There will no longer be any *here*'.

The first key point is that this latter thought – no longer any *here* – is first-personally unthinkable. Anyone who doubts this is simply invited to try thinking it. He will fail because he will be trying to conceive of a future state of affairs that consists solely in there not existing something that is for him a necessary correlative of *any* state of affairs. As the nineteenth-century German idealist philosopher Arthur Schopenhauer puts it in making this point, 'the attempt to think … the conditioned without the condition, the supported without the support, fails every time, much in the same way as the attempt fails to conceive an equilateral right-angled triangle'.[12]

Why this impossibility? Subjectivity, as such, is the necessary correlative of states of affairs because nothing can emerge to distinctiveness as any sort of state except as present to an attention that distinguishes it, just as nothing can be foreground rather than background *in itself*. (More fundamentally, nothing can be an object, that is for example large or small, nor indeed any size, in itself and apart from the comparison with other things that require that it be *attended to*; the subjectivity of attention and the object-hood of what is present to it are really just two aspects of the one single idea.) But then, any state of affairs, to exist as such, must be conceived as attended to – and that means, attended to *as from here*. That claim seems at first blush to be a solipsistic absurdity: as if the world, the totality of states of affairs, were dependent on my existing to conceive of it. Yet there will certainly remain after my death, as there already were before my birth, other subjects, each equally attending from his or her own 'here'. But once again, if we press on this notion of 'other *heres*', we find that while it makes perfectly good spatial sense, it won't do the job for which we want it in the present context. This point in space is here-to-me, and that one over there is here-to-you – that is unproblematic, and your *here* is a point that I can readily imagine occupying. But the sense of *here* in which my perception and action, wherever I happen to be located, are *from here*, isn't like that: no one else can occupy it, and by the same token not only can I not occupy anyone else's, I can't even *conceive of* occupying it. I cannot finally look through anyone else's eyes, though I can imagine (that is, bring before my own mental eye) what the view through their eyes might look like to me. Similarly, I can recognise the necessity of subjectivity in general as the necessary correlative of states of affairs, but I can conceive of what that involves only in terms of *how it is from here*. I can only think the subjectivity or 'from-hereness' which is the precondition of any state-of-affairs, as *mine*.

That means that when I pursue some apparently quite ordinary and plausible reflection on my own mortality such as: 'After I am dead, the wild daffodils in the paddock will still come up each Spring', and think of their thus remaining as a certain kind of object for others, I am always covertly smuggling myself back into existence as the subjective ground of that objecthood. I am always, in thinking of a world persisting after my death, to this extent like the runaway lads in *Tom Sawyer* who creep back into the church gallery to enjoy watching their own funeral service.

But of course, matters can't actually be like that: and this is the second key point. The ordinary physical world which is, equally, the necessary objective correlative of attention, is *real* – it is third-personally graspable as subsisting intersubjectively between me and all the other subjects whose existence I can't help but acknowledge. The very idea of its external existence involves the understanding that it is governed by exceptionless causal laws: for just as the world is not my mental construct, it is not anyone else's either, and it can only exist thus independently to be thought about in virtue of being wholly structured by rational regularities that no one can *think* away. In this objective phenomenal world, I routinely both hear of and see what happens to a variety of living things, including other human beings, at their death. I know that this involves the rapid disintegration of all the bodily organs, in particular the brain, that support consciousness in those organisms enjoying it, and in humans the reflexively conscious self-awareness which must underpin subjectivity. Since my own bodily organism can be no exception to the remorseless causality of these processes, I know for certain that the underpinning of my own conscious subjectivity must similarly fail, in due time.

Here is exposed clearly the radical division in our self-knowledge that leads us towards Larkin's intimation of endless anaesthesia. From the perspective of first-personal consciousness, I cannot conceive of my own death because I cannot think of the abolition of the subjectivity which that consciousness essentially comprises along with the persistence of any objective world in which *I will have died*. But knowing what we now do about the brain-based nature of consciousness, the inconceivability of this cannot imply thinking of my conscious self as immortal, since one thing I know perfectly well, thinking third-personally, is that as material existent I shall die, and the organic substratum of my consciousness will be dissipated. Nor, equally, can it lead to the solipsism of thinking that the world will end with me – for why should my death be any different in this essential respect from that of others whom I and the objective world have already survived? I cannot, that is, as inescapably a subject of both first-personal *and* third-personal thought, seem to make any integrated sense of my own oncoming death.

The image of permanent anaesthesia expresses this inability as the prospect of a kind of eternal suspended animation – a state of permanently frozen but somehow still unextinguished possibility. A complementary visual image would perhaps be Damien Hirst's pickled shark. (And probably more people can call to mind this famous, or notorious, piece of conceptual art than will remember its title: *The Physical Impossibility of Death in the Mind of Someone Living*.)

Mortality, finitude and significance

This is certainly a special way of thinking about what is in store for me. But we have not yet explained why it should be, as Larkin calls it, 'a special way of being *afraid*'. Isn't fear irrational at this level in its turn? As to incomprehensibility, 'I know this subjectivity must end, but I simply cannot conceive of its doing so' indeed records bafflement, but why should it inspire fear? I simply cannot conceive of an equilateral right-angled triangle, but this inability doesn't tend to scare me. And it is not as if I am going to *feel* suffocatingly immobilised in permanent suspension, because (like the shark, in fact) I won't be feeling anything. To repeat: the fear to which the prospect of my death gives rise cannot be an experiential or phenomenal fear, a fear of 'what it will be like', since the fearful thing is that it won't be like anything.

Therefore the fear of its not being like anything must be what we might contrastingly call a noumenal fear: fear aroused by not-like-anythingness as such. What kind of grip can we get on that idea?

We could start from the evident fact that when we wonder, as we can't seem to help doing, 'What will it be like to vanish into nothingness?', the obvious answer that it won't be *like* anything cuts no ice – we simply go on to wonder what its not being like anything will be like ... This failure of the obvious answer is sufficiently *explained* by the way non-existence necessarily resists subjective conceivability, as we have seen – but the explanation seems to do nothing to abate the pressure of the question: we can't help being driven to push our imaginations in this regard further than we know they can take us.

Perhaps therefore this insistent question should be understood as asking: what does oncoming nothingness *mean* for my subjective life? How can I reconcile the prospect of subjective nothingness with the *significance* in terms of which I am always grasping my experienced life? Trying to imagine the present or prospective absence of consciousness just as such is evidently futile and self-defeating. But imagining the loss, with death, of the significance in terms of which our kind of consciousness is always realising and structuring its experienced life, is terrifyingly possible. Here again is 'the anaesthetic from which none come round'. I cannot think 'There will be no *here*', as we have recognised. I can, however, think 'Nothing will *count* here' – I can think 'here' as empty ('no sight, no sound') and thereby as absolutely void of significance ('Nothing to love or link with'). But the oncoming of an inescapable significance-void evacuates significance *now*. The real terror of looming post-mortem blankness springs not from 'what it will be like' to experience, but from what it is already like to experience the dissolution of life-significance which the prospect that it won't be like anything brings with it.

Only a little reflection is called for to see this dissolution happening. For instance, I very much hope (at this moment, as the writing proceeds) not to have to die before I have managed to finish this book. Why does that matter to me? Well, for one thing I should otherwise die with a pang of regret (at least in that particular regard), rather than in the glow of achievement coming from having at least got said what I have to say, however imperfectly and however few people end up reading it. But then, why would *that* matter to me – since either pang or glow will be swiftly over, and then, nothing ... There will be absolutely no difference for me, after death, between having succeeded and having failed in this writing, nor yet between having been concerned about success or failure and having been entirely indifferent to them. And clearly, that recognition immediately extends itself to any other enterprise I might engage in, and thus to the totality of my enterprises. I shall be just as dead, just as little bothered and just as little bothered about not being bothered, after a life of effort, self-denial and (maybe) achievement, as I should be after a life spent giving up spinelessly whenever the going got tough. But in the light of my certain knowledge that this is how it will be in no very long time with anything's bothering me, the question 'So why bother *now?*' at once threatens to become unanswerable.

What keeps that question at bay, of course, is normally my constant busy engagement in making day-to-day sense of my life in the present. But that engagement isn't really robust enough in face of the challenge, and for a reason that goes to the heart of what is at issue here. If I ask what sense I am making of my life, that is to ask things such as: what am I *achieving*? (what are my projects, by way of artefacts or ideas or organisation or …?); what am I *learning*? (how am I evolving and developing as a maker of ongoing sense?); what are my *loyalties*? (what commitments structure my shaping and learning activities, what and who do I value intrinsically?). The awareness of these and similar elements in relation to one another, a relation always resolving itself onward as I live from day to day, forms the inhabited significance of my life. And it is the ongoing inhabiting of significance that gives my life meaning or purpose. This is not to be thought of as the pursuit of some overall 'life-project' (though the pursuit of particular projects importantly contributes) but as a specific kind of life-fulfilment of the kind of creature we are: life is meaningful (purposive) insofar as I am working out the ongoing relations of these core elements (which among themselves of course set up subsidiary norms of intelligence, effort, persistence and so on) in accordance with the living normative drive to make the *best* sense I can, the richest coherence, out of my life-experience.

Now each of these key concepts for life-meaning in its different way envisages a *completable* life as its context. Consider achievement. In normal circumstances this doesn't just mean successfully doing any old thing – tying your shoe-laces, washing the dishes, walking the dog … – but succeeding in something that is in some way central, something that carries weight in a life or (to vary the metaphor) properly occupies a fair bit of life-space. But this, of course, it can only do if life-space is finite – nothing can take up either a large or an insignificant proportion of an unbounded area. One's life-significant achievements are those that loom sufficiently large within a view of one's *life as a whole*. Again, take learning. While its specific demands can be as open-ended as the challenges with which one has to cope, one is only *learning* rather than simply adapting oneself ongoingly if one is also cumulatively developing oneself as a sense-maker. But while change can be open-ended, development can't: it implies in principle a life-totality in relation to which one is at any point more or less developed. Or again, loyalty: one's deep commitments are always implicitly made with a whole life in view, since loyalty pledged until the next emergent review-point in an open-ended process of shifting life-circumstances isn't *loyalty* at all.[13]

But from the first-personal, subjective point of view, my life simply can't be thought of as completable, since trying to think its end involves trying to think something coming after the end, and this must project me subjectively beyond any ending. The peculiar blankness with which my death confronts me is really the same thought-resistance as you encounter if you try to think of *the end of time*. (Your thought is rebuffed because all ending is necessarily *in* time.) But this is also, by the same stroke as it were, to think the necessary endlessness of my engagement in my own life. So I cannot escape for long from the question insistently posed by ultimate loss of significance, simply by focussing on proximate significance. My day-to-day busyness, to be found meaningful, must propose its context as that of a whole, completable life – and trying to think that wholeness takes me straight back to the thought-baffling endlessness presented by the prospect of my death.

Let me press this point again through the personal example, since it is crucial to the argument. This book, my current project, seems to matter to me as a potential achievement: I hope that it will express a deeper insight than I had managed to attain in previous work. It matters personally, that is, at least as a stage in my life-effort. But it is precisely that sense of achievement within the context of a whole life that is dissolved by my inability to think the finitude of my own life. A process whose ending cannot be made sense of, is a process whose not-yet-having-ended is equally unintelligible. I cannot make real to myself, from within, the idea of a to-be-completed life in which this project, in order to matter to me in that way, would have to take its place; if I try, any prospective wholeness vanishes from view as we have seen, leaving me with an appalled recognition of its being equally senseless that I am *here*, that I have reached to where I have. Since it is thus senseless that the book matters now, it cannot really matter after all. Here is the very nerve of the 'Why bother?' question exposed to view.

Thinking third-personally?

But isn't thinking about things in this way to restrict oneself quite gratuitously to the subjective, first-personal perspective? Surely the prospect of one's death can be differently approached? 'When I am dead, it will be as if I hadn't been' can be recognised as simply a mistake, which we can expose by asking: to whom? It won't be thus to me, since nothing will be any way to me, and that is not the same thing as everything being the same way. But nor will it be so to others, on many of whom I am bound to have impacted in some way for good or ill, and on at least some of whom I may well have impacted significantly. By the same token, it seems that my life can not only matter (thus answering the 'why bother?' question) but matter as a completed whole, to the *third-personal* view. And while others are bound to take this view of me, it is also open to me to take it myself, at any rate periodically and as necessary – and that is, in particular, when this sort of issue of life-significance comes up.

At stake here is the possibility of the thought developed by the German existentialist philosopher Martin Heidegger, that life is 'being towards death'. According to a recent account, for Heidegger 'Death is a constant shadow accompanying every action, even if we never think about it … understanding death is a demand to *understand life as finite*'. This means

> having a coherent grasp of one's full temporal existence – past (birth), present and future (death) … Heidegger seems to say that our awareness of death should be a constant accompaniment to any living moment … Appreciating our finitude enables us to live each moment with full appreciation of its distinctiveness and the weight of responsibility implied by this view.[14]

He also claims that 'when we anticipate death it frees us, because death illuminates all other possibilities as being part of a finite structure. Viewing ourselves as such a finite structure enables us to view our existence as a limited whole.' And if seeing my life as a finite whole is thinking its completability, I can only do this explicitly by thinking third-personally about the significance of my whole life – that is, by adopting for this purpose the perspective of someone else.

But this isn't actually possible. That I can only think an objective world from the 'here' of my continuing subjectivity is a formal constraint on *conceiving* myself as extinguished by death, which doesn't stop me recognising, in the abstract, that there *will be* a world post-dating my life, or wondering intelligibly about this world at the propositional level. (Will London be underwater by the year 2100, for instance?) I can think about the world of facts for others in a plausibly third-personal way. But thinking about significance for others, before or after my death, does depend essentially on taking myself to be, or reintroducing myself in thought as, the ultimate ground of sense-making. This ultimate ground must be first-personal: what others find significant is only so, insofar as I can make sense of it as significant.

The world of facts can only be imagined as *my* world, but can be thought of third-personally as *the* world; the envalued world, however, the world as significant, has to be conceived first-personally as my world or the values vanish (or remain only as socio-psychological descriptions, which is the same thing). The allegedly third-personal thought that 'My book will embody truth for others after I am dead' represents my book as mattering after my death, only insofar as embodying truth is something that *I now* take to matter. (By contrast, 'Copies of my book will still be around after I am dead' can perfectly well be thought third-personally, even though in imagining anyone then reading it I have to be sneaking back in to peek first-personally.)

So I can't think of my life first-personally as completable; and while I can think of it third-personally as completable, I can't think of it third-personally as mattering. That is, as long as the first-personal and third-personal are our only options for thinking significance, I cannot think the completability of my life *as inhabited significance*. This is the metaphysical fear, indeed the existential dread, with which the prospect of our mortality confronts us under present conditions. My death threatens me with permanent unanswerability of the life-question 'What, ultimately, *for?*' – and if that question is not in some sense answerable, the point seems to drain out of all of life. While our multiple everyday activities can have point insofar as their completion is at least possible, activities to the point of which there is no overall point, in terms of a whole completable life, can only seem ultimately pointless. Full recognition of this, indeed, represents the normalisation of a state of mind classically associated with Macbeth *in extremis* – the anguished perception of life as

> … a tale
> Told by an idiot, full of sound and fury
> Signifying nothing.

And there is more than one kind of idiot: a tale told by one needn't be a matter merely of ranting gibberish. Persistently presupposing an ending but never reaching one would equally be idiocy.

Nor can confronting that recognition be deferred to some future date, sufficiently distant (I might hope) that I can ordinarily avoid thinking about it. 'The time of death is every moment', in Eliot's terrifyingly laconic phrase.[15] At one level this is a reminder of our permanent liability to be hit by the unpredictable coronary or falling tree; but more deeply, it is about the existential reality of our condition. Just as no one who expects life after death will ever be disappointed, so no one has ever experienced his ordinary confident trust in living beyond the present moment as having been misplaced – but then, since if *this* moment is my last moment I shall never know that it was, for all I ever do know it might as well be, and suddenly you are on the edge looking over, or perhaps more accurately, aware of a chasm miles deep beneath the floorboards. And at the time of death which is every moment, I remain incomplete. Every moment, reflected *in*, as well as *on*, sets the significance-dissolving seal of incompletability on our lives.

Progressivism: the project of having projects

Our fear of death, then, is at bottom our finding unendurable the void of significance with which we are confronted by our inability to make any integrated sense of finitude in our lives. Progressivism is then such a powerful temptation because it seems to offer us an escape route from this fear. The escape lies in accepting *in*finitude as, precisely, the underlying significance that our lives should be taken to have. It is our presenting ourselves, collectively, insistently and systematically, with the *project of having projects*: something in which success *consists* in permanent incompleteness. The force of our recoil from meaninglessness is such that, as we have seen, mainstream environmentalism has in fact been turned into an ecologised form of this progressivist response – environmentalism, which of all potential counter-forces has been the best informed and should have been the most compelling.

As a prelude to thinking what might be done about this – or, at least, how we might hope still to retrieve something from it – we really need to understand the deep energy-source of a habit of mind that can appear to swallow up its epochal opposition in this way.

Let me stress *habit of mind*, in order to get beyond the rather routine response that here points the finger at *capitalism*. It is true that capitalism has a history of neutralising and then absorbing forms of opposition to it, from Lloyd-George populism through corporate-state-ism to democratic socialism and latterly communism. But it has been so remarkably resilient in doing this precisely because all these prima facie challenges have shared with it the underlying structural features of progressivism: most obviously, the commitment to constant prospective material betterment ('jam tomorrow', as C.P. Snow notoriously dubbed it[16]) as the central human purpose and source of life-significance, and the accompanying willed optimism about human improvability. The point now is that environmentalism too, which ought at last to have proposed different premises, turns out to have been captured by these same structural assumptions. This should be seen as a very important datum for determining the real culprit here. That capitalism appears to be the default political mode of progressivism is certainly noteworthy, and interesting speculations about human nature and its political expression have been founded on noting it.[17] But it is *progressivism as such* that we have to understand; and its energy comes, as I have been suggesting, from so deep a source as to be metaphysical in the sense of that term that I tried to capture earlier. Understanding it is correspondingly a *conceptual* challenge.

Structurally, the project of having projects is the only kind of life-activity that answers to the radical disconnection between our first-personal and third-personal perspectives. It is the only project in pursuit of which we can permanently achieve through the permanent elusiveness of achievement. That very elusiveness of fulfilment is turned by progressivism into the only kind of achievement that necessarily doesn't elude us. This pursuit alone succeeds (we have projects, which was our project) and yet has its point through the inherent impossibility of success (our project is to need always more projects).

The first beginnings of this life-pattern, which has since come to assume so central a role in Western culture, are (I suspect) what the great seventeenth-century English philosopher Thomas Hobbes was pointing to when he wrote that:

> Continuall successe in obtaining those things which a man from time to time desireth, that is to say, continuall prospering, is that men call FELICITY; I mean the Felicity of this life. For there is no such thing as Perpetuall Tranquillity of mind ... because Life it selfe is but Motion, and can never be without Desire, nor without Feare, no more than without Sense.[18]

But Hobbes lacked the existential perspective to recognise that this necessity of 'continuall successe' arises not just out of restless organic urgency, but out of the very conditions of subjectivity itself. Accordingly, he also thought that the inherent instability of the project of pursuing 'continuall prospering' was a matter of the interpersonal conflicts that it was likely to generate, and that it could in some measure be obviated by ensuring a *political* form of stability. The real instability, however, goes much deeper, and the fully-developed project in contemporary conditions exhibits it with more clarity than could ever have been available to Hobbes.

Once this is recognised as a project of *significance*, indeed, rather than simply of economic and political organisation, the instability becomes obvious. To have goals and purposes just in order to give the business of having goals something to succeed in, is very swiftly going to be recognised as futile.

This is why our mundane life-*practices* of significance, as we move through them, always onwards to the next job, partner, article, upgrade, holiday…, can't answer just at that level; without some relation to an overarching *narrative* of significance, their constant succession will rapidly come to seem just one damn thing after another. So one must be able to justify them by reference to how they subserve higher-order goals – not all of them all the time, of course: lots of our purposes are just matters of private random gratification, and rightly so. But our more structural purposes, at work or in child-rearing for instance, will always tend to have some implicit reference to some more general, longer-term goals. And these in turn will require reference to still higher-order goals, which if the regress isn't to continue must eventually be able to appear obviously self-justifying despite being permanently unachievable. *Making everything indefinitely better for everyone* is the natural candidate for such a goal – indeed, it is the only such broad ambition that meets the case.

The utilitarian mindset evidently represents a key resource here, just as its full emergence in the nineteenth century (so forcefully, that we have never really got free of it since) represented a key transition in the development of progressivism. If our obligation is fundamentally to increase the overall *quantity* of something (whether characterised as 'happiness' or in some other way), we have a project of value-pursuit that can never fail us, just as the project of counting as high as we can go can never fail us. Incompletability cannot rob life of a point or meaning-giving goal into which incompletability has itself been turned.

For all that, progressivism will, of its nature, remain so inherently unstable that it will always need to be reinforced by various forms of cultural lock-in. Of these, what we might call the 'knee-jerk ethical' is the most obvious. Recall Tim Jackson's exordium with which we began the chapter: 'the elimination of hunger and homelessness, and end to poverty and injustice … A better society for our children.' Evidently no one decent could fail to want this latter, but then by the same token nor could our children, grown up, fail to want it for theirs (nor we to want them thus to want it), and so on and on – the proleptic moral impetus is easily established. It becomes progressivist when it is assimilated, in that very characteristic slide, to the material conditions of life. The elimination of hunger (for instance) would be an important social improvement, and we want a better society for our children and on their behalf for theirs … so plainly we are committed to working ('progressively') for, *inter alia*, the elimination of hunger (which we know at the same time, however, to be an unachievable goal in any imaginable practice).[19] The dynamic of perpetual material betterment is thus morally endorsed with hardly any intervention from actual *thought* being required. All the uncomfortable other-things-being-equal clauses that demand laceratingly hard judgements are readily suppressed: we don't ask whether the attempt to eliminate hunger for an ever-growing population by either grossly overtaxing the Earth's resources or taking huge techno-heroic risks with the biosphere *does* represent social improvement, or whether a better society for our children might not actually be one in which Malthusian elimination operated. Or, if this question is allowed briefly to cross our mental screens, willed optimism (so easily consonant with the illogic of progressivism, as is here apparent) is called forth to dismiss it as morally intolerable, and so not a real question.

Cultural lock-in also of course works through strong social pressures that have nothing to do with ethics. We can cite again the paradigmatic Jackson, as already quoted in Chapter 2: 'stuff is not just stuff ... Material things offer the ability to facilitate our participation in the life of society'. Often enough, in fact, they don't so much offer to facilitate our participation as *compel* it – planned obsolescence of so many consumer 'durables' being a characteristic case in point. Even at levels where we appear to have more choice, however, participation is largely imposed on everyone lacking the disposition to become a hermit; there are tacit progressivist standards for everything from our charitable giving through the frequency of our holidays to our appearance and personal hygiene, policed with silent relentlessness by the tyranny of the majority. Our way of life is so set up (and, as all experience shows, any plausible non-capitalist alternative would soon enough also be so set up) that the stream of ever-renewed aspiration towards perpetual betterment, as expressed in both material and non-material terms, just keeps rolling along (like the Mississippi) with a compulsive sweep vastly more powerful than any merely individual doubt or protest. The awareness that it is going nowhere – that the project of having projects is essentially empty – has, comparatively, no strength to prevent our being carried on down with the stream.

While there are indeed these powerful cultural forces operating, the decisive intellectual context for the emergence of progressivism, and the really fundamental form of lock-in, has been *scientism*. This is not just the now firmly embedded belief that science can answer, or at least meaningfully address itself to, any real question. Underlying that belief, and supporting it in a way that enables those persuaded by it to ignore the counter-instances confronting us in almost every domain of actual living, is a commitment to thinking of the phenomenal or natural world that science systematises, as *co-extensive with reality*. This is the root idea that the world available in principle to scientific knowledge comprises (in a phrase of the late English philosopher Bernard Williams) 'what there is *anyway*'.[20] This belief-commitment leaves human beings only two kinds of thing to be – corresponding to the two fundamental and radically distinct elements out of which the world of experience is constituted, its subjective and objective 'poles of construction'. We can be what we are for science, organic entities fully subject to the laws of physics, chemistry and biochemistry – an especially complex kind of living thing. Or we can be Locke's 'selves to ourselves', self-presented consciousnesses each constituting a necessarily unshareable perspective and uniquely privileged over its own exercises of will and of awareness. The deep problem with this dual-aspect nature of our being is the one that our awareness of oncoming death exhibits to us – the problem that makes it an uninhabitable or *unliveable* self-conception in a way that the problems of freedom of will or of consciousness don't (or at least, don't on the surface), and which prompts the progressivist recoil.

Progressivism is thus the overarching form of human life's trying to grasp itself as meaningful, in the only form that the essential incompleteness of subjectivity confronting mortality can compass – this being, for significance, the only perspective on our experience that scientism leaves open to us. Progressivism is also, and as a natural consequence of that dynamic, the soul-sickness that, armed with modern technology, has trashed and is trashing the Earth. It is soul-*sickness* because it can only really be pursued in hollow-heartedness and hysteria – that is, under the necessary self-intensification of bad faith – and at bottom we know this. We are well enough aware by now of what the various goals and sub-goals of human improvability are doing both to the biosphere and to ourselves. We also recognise, tacitly, that commitment to them is a way of trying to escape a fundamental existential dilemma. But that escapist momentum can't be admitted, if progressivism is still to go on doing its job; and yet the non-performance of such a function is unendurable – it leaves us facing permanent incompleteness and the draining away of life-significance. So we intensify our pursuit of 'progress', suspect the intensification and again intensify. And hence, given time and technical reinforcement, we reach by a remorseless logic the literal insanity of pursuing anything remotely in touch with a Western 'standard of living' for all potentially nine billion inhabitants of a planet, the material and ecological resources of which, it is already starkly evident, cannot support that style of life even for those who already have it. Progressivism is able to insist on this mad logic because it represents human mortal subjectivity licensed by scientism and out of living control – spreading itself into all our life-conditions like a cancer.

Notes

1 All the quotations here are from Gray (2004), pp. 2–5. See also Gray (2002).
2 Cited in Gray (2004), p. 17.
3 Jackson (2009), pp. 1–2.
4 Gray (2004), p. 23.
5 Batchelor (1997), p. 34.
6 See the essay 'On Death' in Bacon (1612/1937).
7 Epicurus, 'Letter to Menoeceus', Book X of Diogenes Laertius *trans.* Hicks (1925).
8 Wittgenstein (1961), para 6.431.
9 Of course, this depends on our discarding the belief that consciousness survives death, a belief that Epicurus lacked and that we must surely now reject as deeply implausible, despite its having enjoyed some popularity in the interim.
10 See the essay 'Death' in Nagel (1979).
11 Larkin (1988), p. 190.
12 Schopenhauer (1819/1958), vol. II, p. 487.
13 As demonstrated classically by the Vicar of Bray in regard to the Hanoverian succession:
For in my Faith, and Loyalty,
I never once will falter,
But George, my lawful king shall be,
Except the Times shou'd alter.
14 The account of Heidegger quoted here comes in Carel (2008), pp. 96–100.
15 From 'The Dry Salvages' in 'Four Quartets', see Eliot (1963).
16 Snow (1959), p. 44.
17 For instance, those of Francis Fukuyama in Fukuyama (1992).
18 Thomas Hobbes, *Leviathan* (multiple editions) Part I, ch. vi.
19 E.g. 'First, the world's 1.3 billion poorest people need to be raised out of extreme poverty. This is critical to reducing global inequality, and to ensuring the wellbeing of all people. It will require increased per capita consumption for this group, allowing improved nutrition and healthcare, and reduction in family size in countries with high fertility rates.' – Royal Society report on *People and the Planet*, April 2012. It is nice to see the Royal Society catching up with the environmental crisis, but this vacuous piece of conventional piety emphasises how automatically the wishful-thinking response cuts in.
20 See Williams (1978), pp. 64ff.

PART II

Hope

4

ENVIRONMENTAL TRAGEDY

In the first part of this book, I have argued that humanity is locked into a trajectory of destruction. Led by the societies and economies of the West and North, but avidly followed by the rest of the world as best (or as worst) it can, the human species is determinedly undercutting the ecological basis for its own continued existence – as well as that of much else within the biosphere. It is pursuing this trajectory most dramatically and comprehensively in the form of anthropogenic climate change, which a sober assessment of the evidence should now persuade us to recognise as irreversible. As we stumble on past successive tipping points and feedback thresholds, this process will produce global socio-economic and geopolitical consequences on a scale ranging unpredictably from the massively disruptive to the catastrophic, and beginning well within the lifetimes of many now alive. Meanwhile the principal agents of these changes, the advanced consumer-capitalist societies, are helpless to halt or reverse them even as they have begun to acknowledge and fear their results, because commitment to the progressivism that drives them has such a firm grip on us. This reflects features of post-religious human experience so general and pervasive as to seem inseparable from our modern self-understanding. Only an extensive armoury of denial mechanisms now enables people in these societies, and their imitators worldwide, to keep at bay the knowledge of what we have done and are doing to ourselves and to the Earth.

That might seem a savagely ironic prelude to entitling the second part of the book *Hope*. And so it would be, if hope had to mean optimism – the belief, against the odds, that somehow all this can still, at well past the eleventh hour, be averted. That belief will only further betray us. But a disillusioned honesty about our situation may remind us that *hope* can mean something much deeper and less defeasible than this. It can mean the clear-eyed determination to live anyway, the life-courage of trusting oneself to the chance, thus kept open beyond anything that either optimism or pessimism could anticipate, that the unforeseen may yet reward us.

Hope of that order, however, cannot consist in mere undauntedness, if we are to live at all steadily by its lights. It demands also real understanding, and at least a scoping of real possibilities. Understanding means not just recognising denial for what it is, but also appreciating what locks us into it and how deep that goes. The business of this chapter is to argue that our situation is genuinely tragic: our environmental plight is not a *problem* that confronts modern humanity, externally as it were, but rather something inherent in the character of modern human being. Chapter 3 has already suggested what is at issue here. In the two chapters following this one, I try to track that character to its roots in the human mode of consciousness and the way our present, historically unprecedented alienation from the natural world has impacted upon it. That is a necessary prelude to canvassing human possibilities that remain open, identifiable from long-standing features of the human condition now temporarily in abeyance. Only in the honest acknowledgement of these features for what they are, can we hope to come fully out of denial and embark on retrieving what is still possible, which is the theme of Part III.

A very important requirement for understanding is thus to grasp the real *shape* or structure of our environmental situation.

We might take that shape to be already plain enough: does it not reside in the pattern of deferred and cumulative externalities, the long-term ecological damage and associated harms to which environmental concern has long been calling attention, and the unjustified burdens that they will impose on future generations? But this is to misrepresent the shape of the issues in closely interrelated ways. The first of these ways regards our ability to perceive this situation as properly tragic. The second regards the dimension of *time*, the shifting of the focus of attention from present to future which has become standard. I shall inevitably have to interweave consideration of both these elements in what follows, though the next two section-headings focus on the tragic and the temporal aspects in turn. I then explore the kind of ethical understanding that can allow our present situation, aside from any future reference, to be genuinely tragic.

The shape of the tragic

If we are pursuing an escapist progressivism, and this progressivism is on course to trash the planet for our successors – what makes that a *tragic*, rather than just a grievous situation for us to be in?

Bad outcomes and conflicts of obligation are not necessarily tragic, even when the outcomes are terribly bad or the conflictedness agonisingly acute. To suppose otherwise is to give in to the casual journalistic misuse of a crucial term. A road accident in which many people are messily killed is appalling in consequentialist terms, but not (thus far) a tragedy, despite the headline-writer's reflex description of it as such. A man who must betray either his friend or his country is caught in a bitter dilemma, but again he is not, just thereby, a tragic agent. But the bind in which we are caught environmentally is tragic in the proper sense of the term.

Again, that is *not* the sense employed by the American political philosopher Stephen Gardiner, for instance, who in his recent book *A Perfect Moral Storm* writes freely of 'the global environmental tragedy', when all he means is that although the relevant facts are now generally known, effective action has hitherto distressingly eluded us.[1] His book is intended to illustrate how this disjuncture between knowledge and action presents us with an ethical challenge, which is nevertheless one of his basic premises that we can still address, if we pull ourselves together.

In its appreciation of any genuinely tragic dimension, this is on a par with telling Othello to get some perspective, or Macbeth a bit more sleep. Much more insight is shown by Clive Hamilton, who as we have already noted in the Prologue calls his *Requiem for a Species*

> the story of a battle within us between the forces that should have caused us to protect the Earth – our capacity to reason and our connection to Nature – and those that in the end have won out – our greed, materialism and alienation from Nature.

This at least introduces the characteristic tragic pattern of inner division, and recognises some of the essential forces in play.

But greed and materialism are not obviously *entailed* by our reasoning powers. Tragedy in the full sense arises when disaster ensues from and expresses destructive weaknesses which are *inherent in the key life-strengths* of an agent, whether an individual, an institution or, as in the present case, a mode of civilisation. This is indeed the kind of pattern that we need to register in the environmental case. Very deep-seated features of the secular and instrumentally rational spirit that has produced so much worthwhile improvement and real progress in the West, have also generated a pervasive inability to rein in the relevant activities before they do irreversible ecological damage. This damage is thus inseparable from distinctive human strengths that Western civilisation in particular has realised – the strengths to develop a sophisticated self-conception, to be rational and make deliberated choices, to base belief on evidence and empirical testing, to free ourselves from ignorance, superstition and dogma. When the very same characteristics, turned pathological, work to blind us to what we are doing and to neutralise any strivings towards self-recognition, all the ingredients of genuine tragedy are in the mix.

The way in which progressivism has played out on the political stage is also tragic. Hamilton illustrates this inadvertently when he claims that 'Climate change represents a failure of modern politics. Elected governments should execute the people's will yet, in this greatest threat to our future, governments around the world have not represented the interests of the people …'.[2] Instead, governments are supposed to have deferred to 'a powerful group of energy companies and the ideology of growth fetishism they embody'. But this blame-game is just a cop-out, as becomes plain when we spot the illicit slide between the people's *will* and their *interests*. The truth is that in this matter 'the people's will' has long been dead set *against* their own interests, and the destructive hold of growth fetishism, as we have seen, goes far deeper than scapegoating the energy companies could begin to suggest. The people themselves have both willed ('democratically') and not willed climate chaos, in the same way that progress has both bettered and radically damaged the human condition.

Tragedy, that is, does not just involve harm occurring through an agent's doing wrong – it does not operate wholly nor even mainly at the level of consequences and obligations. It involves instead, we might say, the agent's *being* wrong, to the extent that all his doing is imperilled, even had the particular trigger for tragic action not been present. This is why tragedy in its literary forms can often appear so gratuitous: if Oedipus hadn't reached that crossroads until a few minutes later, if Cordelia had been even ordinarily tactful … There is an important sense in which what actually happens in such interactions matters much less than the character of being that it brings into play. What specifically *occurs* just realises the underlying character of certain kinds of flawed strength.

This is a dimension in which we habitually misrepresent our environmental situation. In claiming that the obvious candidate for a key strength which is also our key defect is the overwhelming orientation towards material progress in Western societies, I am not after all saying anything very startling. It is far from a novel observation that progress in this sense has brought signal benefits, not just in the West but worldwide, and exhibits the dynamic, innovative, constructive, knowledge-based nature of Western civilisation since the seventeenth century, but has also so heavily impacted the biosphere as to have jeopardised the habitability of the planet. Recognition of this double aspect, however, now very typically goes with attempts to sketch out how we could maintain the benefits while avoiding the harms. Such attempts have lately become reasonably common currency among the environmentally aware – I have already mentioned Tim Jackson's *Prosperity without Growth* as an excellent recent example of the genre. I don't think, however, that we can understand what is really wrong with progressivism – how strength and destructive weakness are tragically interwoven in it – with this meliorist cast of mind, which both furthers and is fostered by denial. As I have tried to suggest in Chapter 3, we have to look much deeper.

One thing should already be clear: if we are talking in this way about the deep structure of motivation, about the underlying radical flaws in an agent's disposition to action, we are talking about the *present* – about how things *are* with the agent, not about how that will or may pan out in any subsequent action. How does that sit with the environmental case?

Environmentalism: the future and the present

Environmentalism has always been about the present at least as much as about the future.

That would go without saying if all it signalled was the recognition that we must act now to prevent damage and harm later. For when else but now, that is, in the present, can we ever act? And how else can we act but with an eye to what will or might follow from doing this as opposed to that – these upshots, in environmental matters, taking typically some considerable time to materialise? I am not, however, seeking to rehearse these unarguable platitudes. What I do want to flag up is no platitude at all, since it is routinely obscured by the now-standard assumption that environmental concern must be 'concern for the future'. It is that this kind of concern is also, and indeed primarily, concern for our state in the present. And not just concern about *the fact that* the present is busy inflicting environmental harm upon the future: rather, a vital component of environmental concern has always been dismay at those characterising features of our way of living *now* – whenever 'now' has been – which express themselves, in our present situation, in that kind of careless stance towards the future.

This distinction is crucial, and it is habitually overlooked in contemporary environmental discussion. Stephen Gardiner, for instance, demonstrates how near one can get to the key point here without seeing it. His book is all about the ethical dimensions of our environmental plight, and specifically of anthropogenic climate change. In pursuit of a fix on the ethics, Gardiner proposes what he calls a 'global test' for our social and political systems. Suppose, he says,

> that human life on this planet were subject to some serious threat. Moreover, suppose that this threat was both caused by human activities, but also preventable by changes in those activities. Add to this that the existing social and political systems had allowed the threat to emerge, and then shown themselves to be incapable of adequately responding to it. Then ask two questions: would such a failure license a criticism of the existing social and political system? If so, how serious a criticism would this be?[3]

He then concludes (flat-footedly enough) that our relevant systems do indeed fail this test, and thus expose themselves to serious criticism as 'inadequate'.

But, inadequate to *what*? If to preventing serious longer-term threats to human life on the planet, which is all Gardiner offers us, then nothing interesting has been added to the hypothesis that sets up the test. He has explicitly built the threatening of preventable harm to future human beings into this hypothesis, and that is already quite sufficient for moral criticism of what the 'existing social and political systems' are *doing*. The questions that he proposes only have any cogency when they are taken to prompt censure of the *way of living* which is doing the threatening. To be placing these life-threatening burdens on the future, we must surely be gravely deficient in human terms in the here-and-now. Gardiner, however, plods on stolidly past this, the only potentially revealing result from his test. Although his book contains some interesting discussions of what he calls moral corruption, the problems of 'lesser-evil' arguments as made by necessarily interested parties, and the general liability to bad faith that the intergenerational situation creates, never in the whole course of it does he discuss what we might call the life-profile of serious inadequacy in the present, except in terms of potential future damage to the biosphere or neglect of future people's legitimate claims and expectations.

But that environmentalism has always been centrally concerned with the life-profile of the human present, whenever that present was, should be clear from the most cursory recollection of its history. As Gary Haq and

Alistair Paul remind us:

> Early environmental campaigning was concerned about rapidly industrialising and expanding cities encroaching on fields, forests and wilderness areas. Groups were formed to protect landscapes for their aesthetic and recreational value and to preserve access to and enjoyment of scenic landscapes as a 'national property, in which every man has a right and interest who has an eye to perceive and a heart to enjoy'.[4]

The embedded quotation is from Wordsworth, emphasising how the lineage of the approach can be traced back to the high Romantic embrace of non-human nature as a source of moral and spiritual regeneration, when so unprecedentedly many had been absorbed into 'cities, where the human heart is sick'. Significantly, the National Trust was founded in 1895 in large part to protect the beautiful remoteness of the English Lake District from Manchester's thirst for water. The same strand of concern is strong in John Muir and the US environmental tradition starting with the Sierra Club and the defence of Yosemite. As I have already noted in the Prologue, this was not, on either side of the Atlantic, about 'the future' in the ecological sense – future generations, human or non-human, whose well-being the anticipated longer-term effects of present impacts on the biosphere were threatening – because the future in that sense had still to be invented. It was, instead, concern about what presently existing people, and their children growing up, needed by way of available wilderness for their own well-being. Muir puts the point in characteristically forthright terms:

> Thousands of tired, nerve-shaken, over-civilised people are beginning to find out that going to the mountains is going home; that wildness is a necessity; and that mountain parks and reservations are useful not only as fountains of timber and irrigating rivers, but as fountains of life.[5]

Recognition that 'wildness is a necessity' continues to be powerfully evident in the much more ecologically-minded Aldo Leopold ('Too much safety seems to yield only danger in the long run'[6]), and even in Rachel Carson's *Silent Spring*, one of the founding documents of modern scientific environmentalism – which is not just about dangerous concentrations of pesticides moving upwards through the food chain, but also very strongly about the *spiritual* travesty of a spring without birdsong. It was along these lines too that Carson the marine biologist and ecologist could write in her posthumously published *The Sense of Wonder*: 'it is a wholesome and necessary thing for us to turn again to the earth and in the contemplation of her beauties to know of wonder and humility'.[7]

We may well tend to find a glance back at this kind of language rather embarrassing, a bit like re-reading one's teenage diaries. 'Beauty, wonder, humility ...' – haven't we grown out of all that self-preoccupied gush about the state of our souls, and at least since the 1980s made environmentalism properly scientific, properly progressive and forward-looking, by turning it into the pursuit of sustainability? But the palpable failure of the sustainability paradigm, if we can bring ourselves to acknowledge it, should give us pause here. Trying to protect the future from the present by leveraging present reluctances with speculative predictions tacitly vulnerable to those very reluctances, has left us facing irreversible damage and possibly catastrophe. Might that not now lead us to look with a fresh eye at these issues?

Sustainable development: shifting the focus forwards

Since it has subsequently been submerged and almost lost to view, it will be as well to give this historically prior cast of environmental concern a broad general characterisation and a working label. We might, then, call it 'right-orientation environmentalism',[8] and sum it up as claiming that the way humans are increasingly living in modern urbanised societies is losing a proper orientation towards the natural world, and needs for present people's own sake to recover it. The 'way of living' in question is always one or another stage in the development of what now, in the early twenty-first century, has become a life lived by the majority in cities. These cities range from huge, by any historical standard, to gargantuan – as the Skidelskys observe, 'an inhabitant of eighteenth-century Paris, then the largest city in the world, had only to walk thirty minutes to find himself in farmland ... his modern equivalent would have to walk six hours through crowded traffic'.[9] Such cities produce and consume vast quantities of non-necessary goods, including food and other products and materials sourced globally (again, with all the transport implications). And they are restless with obsessive mechanised mobility. They exhibit all the symptoms, in fact, of what we have earlier registered as vacuous unease.

Originally, then, environmental concern is 'right-orientation environmentalism' addressed to what is amiss with that kind of civilisation in the given present. *One* aspect of what is amiss, it then begins to emerge, as understanding of ecological interconnectedness and its implications increases, is our being now also on course through the consequences of these life-patterns to wreck the planet for future habitation a good way down the line. That such concerns find themselves subsumed, as the twentieth century advances, into a discourse where what is amiss with us is principally *that we are* wrecking the planet 'for the future', is a crucially significant part of this whole history. It is the story, in fact, of how the 'sustainable development' model already to be glimpsed in the 1972 *Blueprint for Survival* has come since the 1980s so completely to predominate in our conception of what environmentalism is about. On this model, the exploitation of global ecological systems to meet human needs has imperilled the stability of these systems as resources for future generations. Given all this, as the way the problem is framed, environmentalism quite naturally comes to be seen as a movement for protecting the human future from the human present. That is what has always been meant by 'saving the planet'. (The planet itself, as James Lovelock enjoys pointing out, needs no saving, since it can and will look after itself, whether or not that involves shrugging off humanity.[10]) The central purpose of environmentalism so conceived has been to identify and promote ways of bringing human–ecological relations back into some sort of working balance, capable of ensuring that a collective life-impetus towards continuing human betterment can be carried forward into the new global context that we have ourselves created.

I have already argued that this sustainable development model is profoundly compromised. Historically, its emergence as the standard discourse resulted from environmentalism's very understandable desire for scientific respectability and mainstream influence, going in one direction, meeting the socio-political mainstream's desire to accommodate this newly unignorable challenge in a form that it could hope to neutralise, coming in the other. As now hegemonic, the discourse's inherent slipperiness plays a central and dangerous role in licensing both the denial that we need major social and political change to address the issues, and the denial that we are now well past being able to prevent much grievous damage even with such change. But the point in this context is, precisely, *hegemony*. The sustainable development insistence on 'concern for the future' as the only genuine kind of environmental concern simply prevents our seeing behind this to the concerns about humanly inadequate present living which environmentalism has always strongly included, and still covertly retains.

The real, practical danger of taking what is environmentally wrong with present ways of living to be exclusively a matter of jeopardising the rights or livelihoods of future people for the sake of present satisfactions, is that the focus then almost inevitably shifts to changes intended to prevent our bringing those consequences about, and away from recognising what the prospect of our doing such harm reveals about our present life-habit – still more, from recognising it as deeply and tragically flawed. But this kind of shift, taking us into a realm of prediction where few quanta are robustly determinate and all specific obligations to future people are down to us to determine, is tacitly set up to let us off any too uncomfortable hook.

Character and action

Seeking to re-embed future-directed environmental apprehension as critical awareness of the life-profile of present civilisation is structurally analogous to a way of thinking about character and action that has recently re-acquired prominence (recently, that is, in the time frame of ethics, which goes back to the ancient Greeks among whom this approach was first mooted). If what makes some morally dubious action *wrong* is only that it produces harmful consequences or that it violates some duty, then we are left unable to find anything wrong with actions that happen (even perhaps by accident) not to produce such consequences or neglect such duties. But we surely want to be able to say that people can produce no harm by their actions (perhaps their malice is always wholly ineffectual) or neglect no duty (they may be coldly punctilious in all such matters), and yet still be subject to critique: to the sort of critique, precisely, that *malice* and *punctilious* here imply.

Another way of putting this is to note that we do not harm or betray one another *at random*. Typically, people who behave like that are the kind of people who *would* behave like that, and to say this is to comment adversely on their habits of mind, tendencies in preference, degree of self-absorption, powers of imagination and a range of other relevant characteristics. Settled carelessness of harmful consequences, or neglect of duty, always evince defects of character and disposition – vices, as the traditional terminology has it, such as selfishness and treacherousness. And correspondingly, of course, settled tendencies to avoid doing harm, to seek others' benefit and to meet one's obligations in the right spirit always evince the kinds of strength of character (kindness, courage, loyalty and the like) that have classically been called the virtues.

There has lately been a debate among philosophers over whether we should go back to regarding this kind of virtue-talk as theoretically fundamental, in place of the grounding of morality in duties or consequences associated classically with Kant and the utilitarians respectively, which has held explanatory sway for at least a couple of centuries. I want to bypass this question in philosophical ethics almost entirely – just now, I simply want to draw attention to the plain fact that there are always these *three* aspects to the rightness or wrongness of any act: not just the benefit or harm that it causes and the rules that it obeys or breaks, but also the kind of agent who would typically do it, the kind of whom such an act would be characteristic.

The relevant point, for the present discussion, is that emphasising the third of these aspects enables us to see an agent as morally evaluable, *particular acts aside*. So someone can tell a one-off lie without thereby evincing or acquiring the vice of dishonesty; it is only if that is the sort of thing she makes a habit of doing that 'dishonest' is the right term for her. By the same token, harm or benefit have to be specifically occasioned, rules have at some choice-point to be broken or actively obeyed – but you can be dishonest (for instance) while you are neither lying nor planning to lie – while you are instead playing chess or making love or doing someone a kindness, or asleep. The vices and virtues are dispositional concepts; that is, they represent ways of being inclined to act, patterns of tendency that we develop into exhibiting. And as an important corollary, how you are in these regards is how you *are* (that is, here and now, in the present), and is not reducible to what you may or may not do, or fail to do, or bring about – whether in ten minutes, or in ten years, or in the lifetime of the next-but-one generation – *because* you are like that.

There is a clear structural parallel between these general features of character in action and the role of 'right orientation' in environmental concern. The parallel brings out that a way of life, an established process of collective agency, can indeed be objectionable insofar as it is actively doing present or prospective harm, or involving us in actively disregarding future people's legitimate claims, but it must also have this third dimension of wrongness. The kind of way of living it is, the character it has as a life-habit which disposes us to do these things, can always in itself be a distinct and legitimate target of critique. The sort of critique that constitutes 'right-orientation environmentalism', in other words, can never cease to be relevant.

There is a further very important dimension to recognising the role of agent character. If we say of someone for instance: 'He is dishonest, because he tells lies', the *because* is essentially semantic: we are saying that 'he is dishonest' means the same as 'he habitually tells lies', that *dishonest* is the word for someone who does this, and so on. But if we say 'She tells lies because she is dishonest', the *because* is, or can be, genuinely causal: there is that in her character and disposition, we are saying, that leads or inclines her to tell lies (perhaps, when it suits her). Agent character is logically not on a level with harm-causing and rule-breaking as features of acts; in an important sense, it is underlying. This comes out clearly enough when we think about the kind of ethical reflection that takes the 'character' route – when we ask, not 'Would doing X count as neglecting a duty?' or 'What would be the consequences …?' but 'What kind of a person must I be, or be on the way to being, that I should contemplate doing X?'. Suppose, for instance, that acknowledging my increasing readiness to contemplate telling lies leads me to recognise myself as becoming dishonest. But dispositions like this don't arise in a vacuum. As one would expect from their growth and lodgement in human character, the virtues and vices are a tangled nexus of mutually dependent and mutually reinforcing tendencies. So asking '*Why* am I tending thus into dishonesty?' might lead me to recognise myself as *cowardly* in certain respects (I typically tell lies when I am scared of the probable consequences of truthfulness). Perhaps this recognition in turn makes it clearer to me that I am really ashamed of certain aspects of myself to which I had supposed myself to have faced up – I can admit these to myself, I now perceive, but not to others, for fear of losing their respect…. Thinking in this way about the character evinced by my dispositions and tendencies to act helps towards a fuller understanding of the patterns of behaviour, the economy of strengths and vulnerabilities, into which behaviour fits.

Now it is very significant for actual moral practice that we have a strong tendency to shy away from this level of analysis when it comes to evaluation of *oneself* – and for obvious reasons. It is very much easier to say 'OK, I broke the rules, but I won't do it again' or 'If I'd known it would cause so much harm, of course I wouldn't have done it', and to feel that we have thereby addressed the given moral delinquency appropriately, than to think down to the murkier, less tidy and much less comfortable question: 'What sort of person must I be to have done *that*?'. It is (superficially) a lot easier to condemn one's deeds, and undertake to alter them, than to condemn and try to alter oneself – although experience shows clearly that unless one does pursue change in oneself, change in one's deeds is unlikely to be very stable. And this reflects, indeed, a further aspect of the way in which agent character is underlying. For if I remain the sort of person who is, at bottom, disposed to lie, I am also going to remain the sort of person who will tend to minimise anticipated harm (to others) and try to massage obligations perceived as inconvenient, when I am considering how to go on in any situation where truth-telling is relevant.

Here the structural parallel with environmental issues is especially telling. It is particularly easy to avoid confronting the kind of life-habit in which we must be bound up, in order to be doing what we are doing to the future, when issues about what we actually *are* doing to the future, and what if any rights of future people we are failing to respect, are as open and negotiable as the sustainable development paradigm makes them. It makes them so because, as I have argued in Chapter 1, there is no genuine constraint exercised by acknowledgement of the harms or the breached obligations in question, so that we need never be made uncomfortable by really *confronting* anything at all. I am not referring here to avoidance of discomfort by ignoring or denying the fact that we are doing environmental harm. We saw in

Part I that there are much subtler forms of denial that can be practised even when the harmfulness of our actions has been conceded. We aren't forced to the point of making this concession in the sort of uncomfortably unambiguous terms – 'wrecking the planet for future generations' – which might force us to confront our habits and their implications, when we can at a pinch hope to call the same patterns of activity 'developing sustainably', and when there are no criteria for doing *that* that are immune to flexing under pressure from our continuing inclination towards the patterns of activity in question. And it is certainly a lot easier to say, of some activity generating CO_2 emissions for instance: OK, we won't do *this* any more (because we think we can find technological ways to do it carbon neutrally, and thus 'sustainably'), than it is to address anything more underlying in our whole technological-manipulative stance. What gets thereby left aside is the essential critique of the kind of civilisation, the character of which is evinced by our threatening the future and that *would still have that character* if this threat could be diminished, or even removed, by the application of smart technology.

Again, we can see the now standard future-oriented structure of environmental and specifically climate change discourse deforming our attention to the issues and misdirecting our understanding of what is centrally at stake – and doing this in the interests not really of the future at all, but of our present need to be able to live reasonably happily with ourselves. It is a way of thinking in which 'progress', if it could be freed from its adverse future effects – if we could pull off, or could have pulled off, the sustainability trick – would be an unalloyed good. The depth at which this assumption is lodged is part of the hold that progress*ivism*, our *pathological* attachment to progress, has over us.

Environmental virtue?

The large majority of philosophers who have addressed environmental issues have done so through an ethical lens. It seems clear from the foregoing that the lens offered by an ethics focussing on virtue is going to be the only one through which we can hope to see genuinely tragic depth in these issues. If attempts to articulate the requirements of environmental responsibility, with respect to climate change or any other problem, were to focus on the kind of character evinced by our present lifestyles, institutions and individual behaviour, rather than (as now quite generally) on the potential consequences of our collective actions or on the extent to which they violate the putative rights of future people, then we might have at least a chance of seeing how this character could involve deep human dispositions that are radically and destructively conflicted.

One thing we might ask is whether, when we criticise our present way of life for being of such a character as, among other things, to entail serious potential harms on future people (and other species), the claim is that we are failing to meet the contemporary requirements of any of the traditionally recognised virtues such as temperance, prudence or charity? Are we, in other words, displaying modern forms of the age-old human defects of greed, selfishness and so on? Are conflicts between these vices and the correlative virtues potentially of an environmentally tragic order? Or is the claim that there is some new, specifically 'environmental' virtue in which we are lacking? If so, how should we describe it, and why might it make more sense of the tragic dimension?

The best discussion of this question that I have come across is by the New Zealand philosopher Rosalind Hursthouse. She argues that when we are trying to pin down what is wrong with us in the present, for us so to threaten and jeopardise the future, a good deal of mileage can still be obtained from well-known and long-recognised patterns of virtues and vices:

> The old familiar vices of pride and vanity make us unwilling to acknowledge our greed, self-indulgence, short-sightedness and lack of compassion; dishonesty, exercised in the form of self-deception, enables us to blind ourselves to relevant facts and arguments, and find excuses for continuing as we are (think of the people who are still pretending that global warming isn't happening); cowardice makes us unwilling to go out on a limb and risk the contempt of our peers by propounding unpopular views, and so on.

> It seems clear that much of what is wrong about our current practices with regard to nature springs from these familiar and ancient human vices – played out ... on an unfamiliar stage ....[11]

– so that *self-indulgence*, for instance, is now exhibited, as it would not have been before we knew what we now know (or at least, should now know), by things such as owning too large a car, wasting electricity, eating factory-farmed meat or products that have smeared a long slick of food-miles across the globe, and all the other aspects of an environmentally casual lifestyle.

Vices such as self-indulgence represent defects of disposition at the individual level, but we can charge our collective way of living (broadly, consumer capitalism) with demonstrating them at a remove, insofar as it is still firmly built on encouraging individuals to develop and display them. And capitalist enterprise itself, of course, is classically subject to some of the more glaring ones. As environmental campaigner Bill McKibben writes

tersely of Exxon CEO Rex Tillerson, a man who claims that global warming is an 'engineering problem' and will not accept what he dubs the 'fear
factor' in connection with it: 'Of course not – if he did accept it, he'd have
to keep his [oil] reserves in the ground. Which would cost him money. It's
not an engineering problem, in other words – it's a greed problem.'[12]

This looks persuasive at first blush, but Hursthouse also canvasses the possibility that we might need to add one or two new, or newly-recognised,
virtues to the traditional list. This would be neither unprecedented nor
implausible, since different civilisations have always subtended somewhat
different sets of character traits as virtues and vices. (Chivalric honour, for
instance, so compelling in mediaeval times, is no longer recognised as a virtue, while Christian modesty and humility would have been regarded by
Aristotle as tantamount to vices.) Ours, as the first civilisation to press up
against and beyond the biosphere's ecological limits, might well by the same
token have turned certain human dispositions hitherto morally irrelevant into
something else. One such candidate new virtue which Hursthouse discusses
is connected with the emotion of *wonder* (and we rightly recall Rachel Carson
here):

> Getting this natural human emotion in harmony with reason really
> matters morally, just as getting the emotions of fear and anger in har-
> mony with reason do. If we think and feel, not that nature is wondrous
> but that Disneyland or the Royal Family of Windsors are, that the
> other animals are not but we are, that the seas are not but swimming
> pools on the twentieth floor of luxury hotels are, and act accordingly,
> then we will act wrongly, just as we do when we fear pain to ourselves
> but not to others ....[13]

The sense of wonder is certainly an environmentally important one. A virtue of appropriate wonder would not, however, be a disposition exclusive
to our relations with the natural world, since it is also arguably called for in
our interactions with great art, and perhaps also in respect of certain capacities of the human spirit to which religion used to speak. Hursthouse goes
on, therefore, to discuss what might claim to be an environmentally specific new virtue which initially (borrowing the phrase from the American
environmental philosopher Paul Taylor[14]) she calls 'respect for nature', but
actually prefers to think of as 'right orientation towards nature'. This is the
virtue that we are trying to inculcate when we teach children not just to
treat living things with respectful awareness of their sentience or natural
patterns of development, but also 'not to slash mindlessly at spiders' webs,
to look at fossils carefully and try to understand their shape, to be glad

rather than sorry that the Grand Canyon is not rimmed with machines dispensing Coca-Cola ...'.[15] It clearly includes something of *wonder*, something of *modesty*, something of *humility*, of a sense of *care*, of *kinship* – a unique admixture, in fact, the effect of developing which is to give one a sense of the inherent worth of a natural world independent of *us*.

It seems, indeed, that we might have to lay the burden of environmental–ethical critique on deficiency in some virtue of this environmentally specific kind, in order to get around an evident problem with relying on the more traditional virtues and vices to support this critique. This problem arises from the tendency of the various recognised virtues to pull in different directions, environmentally speaking. Maybe it is self-indulgent, as things now are, for Westerners to binge on a superfluity of consumer goods, and greedy for oil executives to think primarily of their profits – but isn't it merely a matter of justice (itself, of course, a major virtue) to ensure the production and distribution of more such goods for those in poorer countries, for whom some of these goods, at any rate, can plausibly be seen as necessaries not luxuries? (Although the biosphere and atmosphere, of course, make no such distinction as to how the detrimental impacts of production and distribution originate.) Again, our technological self-confidence may well be a form of arrogance: but surely failure to try deploying it when widespread desertification and consequent famine are the alternatives would be to fail in both compassion and charity? (This is Mark Lynas's argument, as we have seen.)

In this difficulty it might seem that recognising a specific virtue of respect for or right orientation to nature (with its corresponding vices) could help. For this virtue – understandably, given the environmental context and ecological basis of everything we do – seems at least in that sense more fundamental than the others that have been mentioned. Hursthouse herself seems to miss this point. She says that 'the (putative) virtue of "being rightly oriented to nature" is but one virtue among many: what one can, morally, do in its name is constrained by other virtues such as justice'.[16] But if, by lacking it, one of the things to which we lay ourselves open is trashing the biosphere, there will soon enough be nowhere left to exercise the virtue of justice or any other. Production pandering to greed in the West, but subserving what looks like mere justice in the South, might then still be subject to condemnation at this more fundamental level, if it nevertheless depends on the relevant kinds of disrespect. In the climate context, in particular, we might come to see that treating the global atmospheric commons as something there for us to exploit and manipulate as far as we can, *even if* we are doing it for what we judge to be humanly just ends, exhibits the essential kinds of disrespect.

It should also be apparent how recognising a virtue of right orientation to nature might make a tragic depth to our environmental situation also recognisable. For the virtues that drive progress – technological adventurousness, a disposition to take creative risks, and in general a preference for ambitious omelettes over unbroken eggs – run deeply counter to the exercise of respectful modesty and humility towards our natural surroundings. And yet lack of those attitudes does seem to rob us of something essential without which a commitment to progress tends readily, and in the ways which we have explored, towards the pathological.

Human flourishing

But now we are beginning to come up against the real difficulty for an understanding of environmental concern as directed in this way at present lifestyles and institutions and their tragic implications. A lurking awareness of this difficulty has certainly helped to motivate the whole 'sustainable development' displacement of critique into the language of rights, consequences and scientistic prediction.

'So what exactly *is* wrong', we can highlight the problem by asking, 'with disrespect for nature in Hursthouse's sense?' If the response is that a settled disposition of that order, combining with a burgeoning global population and humanity's ever-increasing technological fire-power, leads to long-term ecological damage and thereby rides roughshod over the rights and claims of future people – well, we have answered the question, but in a way that lays the ethical weight once again firmly on what we anticipate will happen in the future as a result of how we are presently behaving. It follows, as has already been noted, that to the extent that we think we can deploy what we take to be a reliable technological armoury to avert these potential consequences, *nothing* will be essentially wrong with how we are presently behaving. It could, in fact, only be an *unsuccessfully* instrumental and manipulative attitude towards the natural world as our environmental context that could ever count as disrespect or wrong orientation on this account. And from there it is an easy step back into a problem-solving mindset and willed optimism effacing any tragic dimension.

Are we not, though, wanting and intending to say significantly more than this when we call the attitudes and dispositions embedded in the Western life-profile disrespectful towards nature? We surely want to imply that an instrumental and manipulative approach here is a profoundly wrong orientation, *even if it appears to work for humans* in terms of its material implications for the human future. We want to say that acting in ways that jeopardise the future environment is severely detrimental to us *now*, and not merely at the comparatively trivial level where, for example, mucking about with the jet-stream starts to produce major UK flooding and rainy Wimbledons – it is detrimental, we want to say, to our well-being in the deepest sense. But this is not just because it jeopardises the future. The life-habit that I described as historically the target of 'right-orientation environmentalism' – increasingly megalopolitan, atomised, addictively consumerist and in perpetual restless motion – is corrupt, defective and humanly inadequate *whether or not* it is fossil-fuel dependent and climate-threatening. It jeopardises the future, in fact, *because* of its being thus inadequate.

But, of course, the whole 'environmental' crux at the political level arises because very many presently-existing people don't seem to feel this, but on the contrary claim to feel that cutting back – even on consumerist excitements, and certainly on the standard of material living to which they have become accustomed – is what would really be detrimental to their well-being. Moreover, many with wider horizons will feel strongly that such cutting-back, with its effects of stopping growth in a now globalised economy, would be even more seriously adverse to the well-being of millions of the presently undernourished and impoverished. This being so, it becomes an urgent question what it actually *means* to say that a way of life that threatens to damage the future deeply damages or impairs us humanly in the present – and, crucially, why that kind of impairment should be more important to avoid than the actual and potential benefits associated with causing the damage are important to pursue.

If we now turn to consider the structurally identical problem that has always afflicted forms of virtue ethics, it should by this point be clear that we are doing more than simply canvassing an analogy. Recall the case of dishonesty. If we say that someone tells lies because she is dishonest, and intend this as a substantive, non-semantic explanation, then what this disposition *consists in* cannot be characterised simply as her tendency to tell lies. ('Her tendency to tell lies is what makes her tell lies' explains precisely as little as the attribution of dormitive power to the sleeping pill explains why it puts you to sleep.) It follows, if we are genuinely to explain anything, that whatever inclines her to tell lies must be something evaluable apart from its characteristic upshot, so that what is wrong with her (in this regard) remains so, even when she is not for the moment telling any lies. It follows too that we must be able to say without essential reference to her lying what we mean by 'wrong with her'.

Virtue ethicists have standardly taken this on the chin, and have offered to supply the required filling-out of 'wrong with her' here by invoking the idea of failing to flourish as a human being. There is scholarly debate over what the ancient Greek word usually translated as 'flourishing' (*eudaimonia*, literally the possession of a good *daimon* or guardian spirit) actually conveyed to the ancient Greeks, but the broad point is clear enough. What is wrong with the dishonest person is that he or she has an underlying attitude to telling the truth that constitutes one particularly-inflected way of having failed to develop properly, and now failing to function properly, as a specific kind of living thing – a human being.

But the attempt to specify rightness and wrongness of character in this way always seems to end up involving an appeal to the idea of human flourishing that threatens to be either implausible, or essentially circular. This is a line of objection that goes back to ancient times, when Plato put a version of it into the mouths of Thrasymachus and Callicles in his dialogues, and it has never been satisfactorily answered. Human flourishing, considered as a condition naturally reflecting our species form of life, cannot plausibly serve to ground the moral virtues of honesty, charity, loyalty, courage and the rest, because its manifestation in any individual human being as a healthy, happy natural life – and how else should a condition of natural flourishing manifest itself? – does not seem to be enough to guarantee that that individual is virtuous. For evidence, one need only look at the glossiness and self-satisfaction of the latest investment banker or cheating 'celebrity' to disgrace the headlines. And equally, people can have all the virtues and still fail to flourish. Indeed in classic cases of self-sacrifice or self-abnegation they can fail to flourish precisely *because* they have the virtues.

Circularity, on the other hand, emerges when we try, in response to that objection, to rule it out by saying that individuals lacking in virtue 'don't *really* flourish'. The wicked, or at least the non-virtuous, may get the power, the fun and the girls, but underneath all that glitz their lives are hollow and unrewarding in human terms, they aren't 'truly happy'. If this were a psychological claim it might or might not be arguable in individual cases. (If we are honest, it seems likely to be true much less often than we, the virtuous, tend to cross our legs and hope for – a point on which Nietzsche traded heavily.) But in any case, it is clear that what is going on here is not empirical psychological observation at all, but the establishment of a claim by definition; human flourishing is being redefined to include observation of at least the important virtues.

The specific form of this general problem with virtue ethics as it applies to the virtue of being 'rightly oriented towards the natural' goes as follows. 'Humans flourish by respecting nature', we say: the 'wonder and humility' comprising a right attitude towards indispensable wildness are 'wholesome and necessary for us', to invoke Rachel Carson again, given our strong tendency towards detachment from the natural in order to manipulate it, and relatedly towards the creation of a ramifyingly artefactual environment for ourselves. But then we have to take into account all the ways in which we have successfully managed and manipulated nature through this detachment, significantly improving our material standards of life over the past three centuries in the process, and in which we might still hope to do so. By the same token, if '*dis*respect for nature' is deficiency in awe, wonder, humility and so on, we can ask what's so wrong with that, considered just in itself? Aren't these an outdated set of feelings about a natural world that we can now analyse so minutely and manipulate so precisely? Now that we can fix genetic codes, release atomic energy and fly to the Moon, isn't it after all human beings that we should be finding wonderful? Perhaps that would be too hubristic; but certainly it would seem that in principle people detached from 'Nature' (as the external, biogeophysical context of human life), and with a manipulative attitude towards it, could still happily flourish in their megalopolitan and consumerist lives, if only they could manage to keep the carbon emissions down and similar detrimental spin-offs in check.

If we then retreat to saying that these ways of alienated megalopolitan living are not *really* forms of human flourishing but manifestations of deep vital impoverishment, we seem to be making the definitional and circularity-introducing move already remarked on. The trouble with that move, however, is that it is wide open to someone (the 'skeptical environmentalist' Bjorn Lomborg, for example,[17] or Mark Lynas in his most recent phase) to respond by dismissing it as merely ideological – in Lomborg's terms, 'the litany'. Define human flourishing in that (outdated, old-green Romantic) way if you like, it could be said – but meanwhile, the large majority will continue to see their flourishing in terms of guaranteeing and extending the favoured lifestyle of the dominant West – and who is to say that they are mistaken, as far as that goes. 'Billions of people want to move to urban areas to achieve increasing prosperity and improve their standard of living', Lynas insists: 'Let us be glad of that.'[18] We should also, on this tack, be glad that we have the technologies to support these life-habits, and that as long as our ingenuity can find smart ways not to cook the planet in the process, continuing human progress can be assured. And with that, we are back where we came in – or at least, where the account given in Part I had left us – with a sustainability discourse and agenda that have patently failed to give us any purchase on accelerating catastrophe, and no other language in which to capture, robustly and decisively, what has evidently gone tragically wrong with the way of life that speeds catastrophe so irreversibly on.

This is the radical bind of which our contemporary environmental plight is the culminating manifestation. It grips us far below the level of the ethical, and largely explains why we have turned with such already disastrous results to a wholly future-oriented form of concern about what modern civilisation is doing to itself. We seem unable to express the rightness or wrongness of a way of life except either in the ambiguous terms of natural flourishing, or via the outward forms, scientistic and legalistic respectively, of consequences and obligations. In a post-religious age we are not only left defenceless as individuals against the existential anxieties that, as we have seen, drive progressivism. We have no language left for saying how that drive expresses something gone wrong, so to speak, in and with our *souls* … at the vital roots of our collective and individual life-ways.

Learning from environmental tragedy

Problems have solutions (if we are lucky). Tragic dilemmas have only tragic (and other) *outcomes*. That is why I have tried throughout this book to avoid the term 'environmental problems', and kindred phrases. Nor, given the wide currency of slipshod misapplications such as Gardiner's, is 'environmental tragedy' actually much better – but I will continue to use that phrase in the sense that I have clarified here.

Tragedy of its nature entails radical breakdown, with terrible loss that cannot be mitigated or compensated. The impossibility of anything that could be called compensation is what always makes the ending of the great tragedies in literature so muted and seemingly flat: as in the final lines of *Hamlet*, given to the burly and utterly pragmatic Fortinbras:

> Such a sight as this
> Becomes the field, but here shows much amiss.
> Go, bid the soldiers shoot.

Words are recognised here as little more than a formal closure that must be gone through, while what has happened must simply be endured. It is also why the eighteenth century, with its neo-classicist insistence on drawing an explicit concluding moral lesson, so that good can always be shown as in some sense coming out of evil, could never rise to the genuinely tragic. In its way stood Milton's influential Hebraicising of Aristotelian catharsis at the end of *Samson Agonistes*:

> His servants He with new acquist
> Of true experience from this great event
> In peace and consolation hath dismissed
> And calm of mind, all passion spent.

But in the great tragedies, our concluding 'calm of mind' is nothing like 'peace and consolation'. It is rather a matter of our having been revealed to ourselves so drastically, and so far below the level of ego and willed self-image, that there is nothing left for the moment to be thought or said.

Yet, however paradoxically, this tragic self-revealing is not incompatible with life-hope, since from it, if we are otherwise sufficiently undamaged, we may still be able to learn. Here, of course, is the rub in the present context. As the reader will already be reflecting, perhaps impatiently, the environmental tragedy is one in which both we and our successors are not spectators at some old play, but ourselves very painfully caught up in the action – and even those who may emerge from it will do so not 'otherwise undamaged', as (in some sense) by art, but grievously harmed in the most material ways. It is clear that whatever geopolitical consequences actually follow from increasing global temperatures over the coming century, any organised life that humans go on being able to lead in the climatically luckier parts of the world is going to be much harsher than anything to which we in the West are currently accustomed. It will not only be far less suavely comfortable in material terms, with rougher hands, more frequently aching muscles and emptier stomachs the norm – it will be much more bleakly stringent morally. Its foundations in compulsion within its borders and harsh discrimination to maintain them will be glaringly evident. Some of those now reading may live long enough to see this, and their children certainly will. To talk of 'learning from environmental tragedy' is surely both callously detached and feebly academic when this kind of future is in prospect.

We must make ourselves do it, for all that. It is, we might think, the price that we must pay for failure, for having missed our epochal chance. (As prices go, it could be held to be a lot less than we deserve.) And actually, the necessary detachment is a measure of how un-academic the insight is. The whole burden of the argument so far is to emphasise that the honesty now required of us demands this difficult, dispassionate stepping-back. Without it, we cannot grasp the real structure of the issues that I have been demonstrating, because unless the structure is recognised as inherently tragic in the way that I have outlined, it is not properly understood, and then nothing can really be retrieved – the equally difficult possibilities of genuine hope, even for a future so grievous, will remain unrevealed.

But, as we have also now seen, properly recognising our environmental plight as tragic requires us to be able to say something more cogent than any form of environmental–ethical or environmental–philosophical thinking has yet managed, about what is radically and tragically flawed in the life-profile of the dominant present civilisation. If the argument of this chapter is correct, what is really wrong with our alienation from wildness and natural wonder isn't to be captured in any terms of future consequences, nor of responsibilities standardly configured as rights and obligations. Nor can it be compellingly expressed in terms of its implications for well-being, at least as that has been conceived by virtue ethics – that is, as the flourishing of individuals within a natural form of life. And yet, pretty obviously, we *are* a natural form of life, an evolved primate species – and nothing environmentally helpful can be said that forgets or downplays this. So if we want to retain the analytical structure that enables us to recognise tragedy here, we are going to have to find some way of showing how respect for wildness is a necessary disposition not just ethically or prudentially, for our well-being, but existentially, for our being at all as human creatures (reflexively conscious organic sense-makers). We are going to have to display this kind of disposition as a natural responsibility *to ourselves*, to be what we fully are. Correspondingly, we shall need to understand how what we are missing when we ignore the insight that wildness is a necessity, and when we live our characteristic lives at such a distance from wildness, matters at such a depth that missing it deforms those lives both individually and collectively – and deforms them to such an extent that the self-defeating escape into tragically destructive progressivism becomes our only recourse.

I do not believe that we can understand these things without exploring the workings of our alienation and the associated deformations at the most fundamental level. We need to ask what it is in the human mode of consciousness, at the root of all our knowledge of the world and of ourselves, that interacts with present conditions of civilisation to generate this failure of natural responsibility. To this question I therefore now turn.

Notes

1 See Gardiner (2011).
2 Hamilton (2010), p. 223.
3 Gardiner (2011), p. 217.
4 See Haq and Paul (2012).
5 Muir (1901/1997), p. 721.
6 Leopold (1949), p. 133.
7 See Carson (1965). I have been unable to establish the page number for this quotation.
8 I owe the prompting for this label to a paper by Rosalind Hursthouse, which I discuss further below.
9 The quotation is from Skidelsky, (2012), pp. 162–3.
10 See most recently Lovelock (2009).
11 Hursthouse (2007), p. 157.

12 Bill McKibben, 'Global Warming's Terrifying New Math', *Rolling Stone*, 2 August 2012.
13 Hursthouse (2007), p. 162.
14 See Taylor (1986).
15 Hursthouse (2007), p. 167.
16 Hursthouse (2007), p. 169.
17 See Lomborg (2001).
18 Lynas (2011), p. 138.

5
STRUCTURING THE SELF

In preparation for the pivotal chapters of this book, I will summarise again to bring the argument up to this point.

I have claimed that contemporary environmentalists are caught in a tragic bind. They are well-placed to see that our newly globalised civilisation is now irreversibly committed to a trajectory into climate jeopardy and massive ecological damage. But with only a few exceptions, their response to this has been reflex denial and willed optimism, kept up with increasing stridency against the clear evidence of facts that they nevertheless increasingly recognise.

This reflex, I have also claimed, is one more manifestation of the genuinely tragic pattern exhibited by our overall environmental situation. Humanity is where it now finds itself – or rather, mostly recoils from finding itself – because powers of understanding, invention and organisation central to the cultural and technical achievements of the West and North over the last three centuries have had destructive weaknesses built integrally into them. The pursuit of progress as real advance, real and historically unprecedented transformation of the material human condition for the better, seems to express the tragic inevitability of progressiv*ism*, which entails the inability to stop this pursuit even when it begins to press unignorably on the limits of what the biosphere can accommodate. This inability has not only generated the slowly gathering environmental crisis itself, but has also intervened to subvert all attempts at a recuperative response. The practical imagining of a green alternative becomes the now mainstream but profoundly compromised mindset of sustainable development.

I have therefore emphasised in contrast to this mindset that our environmental troubles have to be recognised as a matter of deficiencies besetting us deeply in the present. They are only secondarily and derivatively (though still importantly) a matter of our adverse impacts on the resource-base and life-conditions of the future. Beyond economics and politics, shaping ideology, working at a level beneath group and even individual psychology, the driving force of progressivism seems to be a refusal of mortality, an inability to make sense of our finitude, expressing in turn a fundamental and radical dividedness in our human being and self-understanding. This dividedness, as we saw in relation to the way oncoming death confronts individual consciousness, has to do with our necessarily having two perspectives on ourselves, the inner or first-personal and the outer or third-personal, which cannot be combined, but neither of which, it seems, we can abandon. And this condition as a ground of environmental tragedy has also somehow to do with the increasing human exile from wildness which environmentalism in its original conception arose to combat.

This chapter and the next are pivotal for the book, because taken together they press the two crucial questions that arise from the foregoing account. First, how *can* there be that kind of radical division in natural beings – in our human selves, as creatures gifted with a terrestrially unique reflexive consciousness, cognitive endowment and practical intelligence, but nevertheless the bearers of an evolved primate life-form? How can anything natural be *radically* at odds with itself? – since the very idea of being a naturally living thing seems to depend on that of autopoietic integration, organic coherence. But, second, even if it makes sense to think that there can indeed be such radical division in a living thing, how could the 'wildness' that Muir and other right-orientation environmentalists recognise as a human necessity help us in that regard?

The second of these questions is evidently not a distraction from central issues raised by environmental concern. But nor, though this may be less immediately apparent, is the first. The ecology of any living thing is a matter of its interaction with its natural conditions: uniquely in the case of humans, that interaction goes on through thought as well as behaviour. 'How *can* we think of ourselves as natural?' is thus not just a conceptual challenge to the core environmentalist contention that humans too have an inescapable ecology – it is also a question to which we must be able to *live* the answer.

These two crucial questions are not meant to be rhetorical, or just wilfully paradoxical. I think that they have answers, and I shall try in the rest of Part II to indicate what I take these answers to be. The intention of this exploration at the conceptual level is to prepare us for considering in Part III of the book how we might actually, in our real present situation, re-encounter genuine wildness, and how such retrieval could help us to respond as hopefully as we can to the grievous damage we are doing and the ecological and climate consequences that are impending.

Facing these conditions clear-sightedly, that is, requires us to be *philosophical*, in more than one sense. We must, for sure, rediscover the stoical, and indeed Stoic, virtues of endurance that go with tragically disillusioned understanding. Only so will we be strong enough to put our hands, without self-deception, to whatever remedial action now remains open to us. But we must also, now, be actively *thinking* our way forward – reconceptualising in advance the habitable human-naturalness which we shall need to repossess if collapse is not to be mere catastrophe unredeemed. Such thinking, however wide a circuit it may seem to be taking away from 'real-world' economic and policy exigencies, is thus pre-eminently practical – far more so, in fact, than the pursuit of allegedly practical solutions that merely presuppose, and thus reintroduce and reconfirm, the framework of understanding responsible for our difficulties. So all I can do is to ask the reader who, in the course of these two chapters, may start to doubt this practicality, to be patient.

Since conceptual work here is inevitable, I have tried in what follows, even more determinedly than elsewhere in the book, to do it at first hand and to engage the general reader collaboratively in the process. This has involved a series of compressions, short-cuts, takings-for-granted and ignorings which will seem fairly breathtaking to professionally philosophical readers, should I have any. But as I have already hinted in the Preface, there is a higher kind of intellectual responsibility than ensuring that one has fully referenced the literature (enormously extensive in the relevant areas as it is) and traced out its recognised debates. I have wanted to write something that would, indeed, strike people familiar with these literatures and debates as cogent. But, absolutely crucially, I have wanted it also to be as accessible as possible to concerned and intelligent lay people. Given our present plight, somebody has to attempt this. It is no doubt arrogant to presume oneself qualified, but as I noted in *The Sustainability Mirage*, risking such a charge is really what writers are for.

Thinking about ourselves and nature

The sense of our radical dividedness goes very deep – as it must do, if it is to threaten us with such a tragic bind as progressivism. This was the theme of Chapter 3, where I was outlining the two perspectives on ourselves, both of which we need in order to conceive of ourselves properly, but which cannot be integrated, in the sense that one's own mortality under one conception, but not under the other, is inconceivable. But I also referred there, in somewhat Cartesian style, to the intimate inward subjectivity, the continuing first-person consciousness that I tend to consider as essentially me. This seems to involve the long-standing dualist idea[1] of a living being divided between its natural organic embodiment and something else that constitutes *itself*.

Thinking about self-division, that is, turns swiftly into thinking about *the self* as such – and such thinking gets us almost inescapably into puzzles that have both a long history and a continuing vivid life in the present. These puzzles remain as contested an arena of philosophical dispute as ever, and also increasingly a matter of something intended to be empirical study, in the field of 'cognitive science'. What I want to foreground here, rather than any particular theoretical line, is what persistently triggers these difficulties – what it is about the human self that is so deeply problematic on reflection that we find ourselves *having* to talk about our being as 'radically divided'.

To explore these matters we will again, and especially in this present chapter, have to be doing metaphysics – which, recall, is only out of the ordinary because it is engaged in tracing the conceptual boundaries of the ordinary and making sense of where they lie. These boundaries are out of the ordinary in the same perfectly familiar, but on reflection paradoxical, way as the baseline is neither *inside* nor *outside* the tennis court. (If it is inside there must be a further, and if it is outside a hither, *different* line marking out the court – neither of which lines exists.) The broad shape of our relation to nature threatens paradox in a way that Chapter 3 has already foreshadowed: we cannot think of ourselves as natural, though we know that we must be. Specifically, we cannot capture our own conscious self-awareness – or at least, what is at the core of it – in naturalistic terms. We cannot include our fundamental form of being in the natural order as we take that to be available to our thought, although the natural order is where we must suppose our being both to originate and to reside.

To make that last point is to emphasise that our default general account of ourselves is now both naturalist and Darwinian. It is naturalist in the sense that has been pithily encapsulated by the philosopher Simon Blackburn: 'To be a naturalist is to see human beings as frail complexes of perishable tissue, and so part of the natural order. It is thus to refuse unexplained appeals to mind or spirit … it is above all to refuse any appeal to a supernatural order.'[2] It is Darwinian, in the sense that any *emergence* over time, of anything from anything within this natural order, must be evolutionarily explicable. Natural selection is the only way in which anything living, including ourselves, develops and grows more complex in the world as we understand it. This secular, naturalistic general picture is now (despite various forms of rearguard religious protest against it) the working world-image of the advanced societies, and is implicit in the terms in which these societies have globalised themselves through science, technology, telecommunications and the capitalist–individualist economic model. It is the human self-recognition, we might say, that keeps the internet functioning.

But for all that, something non-natural, or *a*-natural, emerges irresistibly *as subject* from within our kind of consciousness, as we exercise it – although we cannot make explicit sense of the emergence of anything *but* the natural from the natural basis that we now, quite generally, take our lives to have.

Consciousness and the self (1): perception

The only way to establish this claim is to show – as far as possible, irresistibly – such paradoxical emergence actually happening. Imagine, then, a simple instance of watching something changing over a brief but surveyable stretch of time – say, a cube of ice dissolving and disappearing in lukewarm water. How is there *consciousness of change* here? Straightforwardly enough, it would seem, I am conscious of the floating ice as being in a certain state (sharp-edged, its original size) at one initial moment, and then of its passing through a series of different states – getting smaller, with progressively more rounded edges and corners, and becoming more and more transparent – over a series of subsequent moments. But something else is required that, focussing on the ice cube, we tend not to notice or think about. For change to be recognised here – for the ice to progress through different states at different time-points over this time-period – what is *doing the recognising*, that is my consciousness, needs to stay the same over the period. For suppose it didn't. Then either a different consciousness would be registering how it was with the ice cube at each moment of attention to it, and nothing would be *recognising change*: each consciousness would just be confronted with its respective ice cube stage, static within the particular moment of attention. (Compare the way the shifting cloud over Helvellyn isn't *seen to change* if one person looks at it from Striding Edge, and another a bit later from Swirral Edge.) Or else, if the attending consciousness changed only at intervals within the period, only a particular phase of the ice's progress from original cube to disappearance would be attended to by any one consciousness, and in respect of *that* phase we should then have to say that the consciousness in question needed to remain the same over the whole phase. (For suppose it didn't … and so on.)

What this makes clear is that the idea of my being aware of the *ice changing*, that is of an entity independent of me undergoing change, depends on the correlative idea of my staying the same.

Someone might respond: all that is needed for *S to change* is that S have property P at time t1 but not later at t2. There is no necessary reference to any fixed point here. And equally, if S is registered by A to have P at t1, and by B to lack P at t2, S will still have changed, in itself and independently of these two consciousnesses. This is true, but it remains the case that for there to be *awareness* of the change, all this will need to be registered by some implicit C whose consciousness will then need to be taken as a fixed perspective. Quite generally, to think *of* change over a period t1–t2, we have to think *as* something that stays the same between t1 and t2.

But as what? What *is* this something that stays the same? I have just now used the term 'consciousness'. But surely that, as we experience it, doesn't remain the same over even the briefest period of time. It is constituted, rather, by a permanent flickering of focus back and forth across a field of attention, a constant intensification and fading, accompanied for the most part by a subliminal undertow of somatic variation and emotional charge that normally forms an unattended background but rises sometimes to notice.

What I have just sketched as consciousness – the ongoing living business of perceptual awareness – is actually, then, another and more intimate process of continuous change of which I am aware, or can become aware in experience. So what must stay the same for us to think of *that* process, must be not consciousness as such but what *deploys* consciousness in both this kind of case and that of external change such as the ice cube dissolving. We must fall back here to the idea of the conscious self (or person) who is attending. For me to be aware of change (and all *awareness* must be of change, or of its corollary stasis, over time), there has to be something unchanging corresponding to the 'I' that is always present in awareness.

But surely *I* am a living being, and as such *do* change over any given period of time. My digestion is bubbling away, my hairs are growing, my cells – including my brain cells, some of which are actively involved in attending (for instance) to the ice melting – are undergoing constant decay and replacement. What this makes plain is that these changes in me as a continuing organism can't make me a different *self*, in the sense in which I must remain the same self to register change in objects to which I am attending. So the self in this sense, it seems, must be something distinct from the person considered as an organic being.

Classically, of course, this distinction has been framed as that between body and mind. Dualist thought all the way back to Descartes has claimed that the *real* I, my real self, is my mind. Anyone with even a little philosophical background is likely to know the sequence of moves that Descartes himself made in seeking to demonstrate this. I am fundamentally, he claimed, what I can't *not* be: that is, what he called a 'thinking thing'. I can't not be *that*, because it is presupposed by any answer to the question 'What am I?', or to any other question, that I can think up – and unlike any other answer, there is no way of supposing it to be wrong, since any supposing I may do, including that supposition, simply confirms it. But something which is fundamentally a thinking thing, whatever else may be involved in its existence as a self, is fundamentally a mind.

But we should now be ready to recognise that the term 'mind' has its own ambiguities in this context. We can make perfectly good sense, for instance, of the idea that my mind can change. I can think one way about something (believing in scientific materialism, liking Tchaikovsky, preferring purple to blue, or white wine to red ...) and later, with more or different experience, come to think differently – come to find that 'my mind has changed' or 'I have changed my mind' about this or that. I can also quite deliberately set out to change my mind, or at least to expose my mind to the possibility of having to be changed. Indeed, as we observed when discussing Mark Lynas' position, being ready sometimes to do that is an indicator of (as it were) personal mental hygiene. If I *am* my mind, however, and yet for me to be aware of anything changing there must also, correlatively, be a *me* that remains the same, how are we to conceive of any of this as possible? Doesn't there have to be something here that is not 'my mind', in any psychologically characterisable sense, but *is me*? When we pursue these thoughts far enough, we do seem to find, as Descartes found, something of this order *inhabiting* our kind of consciousness – looking out from it, or through it.

Consciousness and the self (2): embodiment

Perception can thus seem to construct the self as a fixed point of reference and, as such, something over against the world of objects perceived. But what about *desiring*? Surely this is an area where the self known as subject of conscious experience can only be grasped in terms that essentially involve the organic naturalness of embodiment? In desiring, I am surely present to myself as something fully within the scope of natural forces. If I want a drink, the drink does not just stand to me as object-for-a-subject in the way it does in simple perception, but as the object of my thirst – it is realised as a motivating goal, not simply contemplated, and I myself am correspondingly realised in this transaction as organically implicated in the relevant part of the world.

Again, my sensations of pleasure and pain, and the shifts between them that are so closely associated with desiring and having my desire satisfied (or not), are processes in which I am involved intimately and organically. And similarly with the logical relation between desiring and acting: if I desire something, I am at least disposed to try to get it, and that must involve being disposed to initiate something organic, at the simplest to move my limbs, but also probably to engage in various vocalisations and other forms of active self-expression.

At the centre subtended by reflexive consciousness, that is, I must think of myself in relation to all my experiences, in every mode of subjective engagement. I am not just that to which my perceptions are referred, but also that on which my memories depend as an entity extended over time, that for which my pleasures and pains are inherently attractive or repellent and my emotions variously compulsive. I am also, as we shall consider further in the next section, that for which my choices are open and to which my actions are attributable. The *my* in all these marks the first-person perspective, the 'from-within-ness', and they are all modes of conscious experience, with a specific 'what-it-is-like'-ness to each of them. At the same time they are all modes of experiencing attributable to a particular organism – modes of the livingness of *this* body, in the unity of which these thoughts share a grounding with this sensation of holding the pen, this feeling of the chair beneath this rump, this residual twinge in the injured foot, this set of internal gastric indications ... I know myself in all these ways as embodied, and everything else through this embodiment: at the core of everything is *me-alive*. As D.H. Lawrence powerfully captures this recognition in his *Fantasia of the Unconscious*: 'Primarily we know, each man, each living creature knows profoundly and satisfactorily and without question that *I am I* ... the vital centre of all things. I am I, the clue to the whole.'[3]

But while this is true, the point is that what Lawrence himself calls 'mental consciousness' (actually, the way our kind of consciousness is *supercharged*, as it were, by our kind of intelligence) *subjectivises* even all this embodied experience, making it intelligible only by reference to an unchanging centre. We can see this in forms of experience such as memory and desire which presuppose an organic embodied dimension in ways that the pure observational focus of perception at least appears not to do.

Consider memory. I can't remember experiencing any event at which I was not personally present, that is as an embodied continuant over time. (If you want to test this, try *remembering* the battle of Waterloo – not when it happened or who won, but actually being there. You can imagine this, if you have read *Vanity Fair* or seen a film about Napoleon, say, but you can't *recall* it.) But this continuant, based on whose continuity I remember things, must be in essence unchanging. Take the most difficult case for this claim:

suppose I now recall a literally life-changing experience that I had yesterday – like the Wedding Guest in *The Ancient Mariner*, who

> went like one that hath been stunned
> And is of sense forlorn;
> A sadder and a wiser man
> He rose the morrow morn.

Still, that happened (let us suppose) to me – I am the same person on the morrow, and must be if *sadder* and *wiser* are to have content. But in what sense the same? I might well be psychologically and intellectually transformed – my values re-valued, my whole understanding of the world revolutionised. I might even be physically different – not just through the ongoing lapse of cells, but as when one 'ages years in a single night', which is something that great mental stress can plausibly do. What then must remain the same can only be me as subject-of-experience, just as in the case of perception in fact.

(An even more dramatic way to illustrate this is in terms of Kafka's famous story. You wake up tomorrow morning to find that you have the body of a gigantic beetle.[4] Since your discovery is that this transformation has happened to *you*, you must still be the same person!)

We can also think in the same way about desire. Am I the same person when I am yearning passionately after something, and once I have had that yearning satisfied? There will be all sorts of psychological and organic differences (due to dopamine release in the brain and so on) between states of desiring and of satisfaction – but desiring only makes sense at all if what I want is that *I*, the person now wanting, will be the recipient of the relevant satisfaction. Thus even if I want to do something – a course of meditation, perhaps – because I think that it will transform me into a wholly different person, the difference between wanting the results of this process for myself and wanting them for someone else remains in full force, which it can only do if what will transform me doesn't alter the central continuing *me* who will experience the transformation.

These deep thematic complications emphasise that the unchanging subjective centre that consciousness of consciousness subtends has to be both *really* and *metaphysically* central – detached from the experience to which it is central, in something of the way in which the central point of a rotating solid disc must be still. Not all the disc can be spinning (around what?), but if it is solid every part of it will be carried round in circular motion with the rest. There cannot be physical stillness at the centre, but unless there is real stillness there the spinning motion can't happen.

But this is not a constructed point like a centre of gravity – a merely theoretical entity posited in calculating forces. The American philosopher of cognitive science Daniel Dennett, who frequently uses the analogy of a centre of gravity for the self, writes that such a centre

> is a mathematical point, not an atom or molecule. The centre of gravity of a length of steel pipe is not made of steel, and indeed is not made of anything. It is a point in space ... a very useful thinking tool ... it averages over all the gravitational attraction between every particle of matter in a thing and every particle of matter on the planet, and tells us that we can boil all that down to two points – the centre of the earth (its centre of gravity) and the centre of gravity of the thing – and calculate the behaviour of the thing under varying conditions.[5]

He thus calls centres of gravity *theorist's fictions*, and the self by extension a *centre of narrative gravity* 'posited in order to unify and make sense of an otherwise bafflingly complex collection of actions, utterances, fidgets, complaints, promises and so forth that make up a person'. A centre of gravity, on this account, and by extension a centre of normative gravity, is no more real than the average British family, with its 1.9 children, is real. The centre of a spinning disc, however, must be motionless, in the way in which only what really exists can be motionless (and not, for instance, in the way the number 3 is motionless). By the same token, the unchanging self subtended at the centre of our reflexive consciousness of embodied thoughts and feelings must be metaphysically real, not just theoretically constructed.

Consciousness and the self (3): the will

Thinking about the reflexivity of consciousness thus introduces us to a centred self that is not simply a notional point. Rather, it expresses itself in a lived 'from-here-ness' that feels fundamentally real. But also part of this feeling of reality is its being a centre of *spontaneity*, where not only my perceiving, remembering and feeling but also my willing and acting take their rise. It is, however, a spontaneity standing over against, rather than belonging in, the world that is the object of either my perception or my will.

Pursuing the project of trying to show this irresistibly, we need first to be clear how our exploration of perceiving change or stasis implies a fully determinate objective world.

If to be aware of the ice cube changing I must *in myself* (whatever that comes to) remain unchanged, then the changes of which I am aware must be taking place wholly within the ice and its relation to its surroundings.

This is evidently the same idea, in a different aspect, as that of the necessarily unchanging subject. It is also, by extension, the idea that what confronts me as subject – or any such subject – must be a world the changing (or temporarily static) conditions of which must be explicable without reference to any particular subject – explicable *in themselves*, as it were. This is, in a word, the idea of objectivity, and we can see how it implies the availability of a third-personal view of the world. If the perspective of my subjectivity is first-personal, its requiring to have objects distinct from it in order to *be* subjectivity must bring with it the permanent possibility, at least, of these being also objects for other subjects, for *him* or *her* as well as for *me*.

By the same token, how it is with these objects must be *wholly* explicable in these objective terms. That is, for anything that happens to or among the objects making up the world, there must be a set of reasons, in terms of how things are with these objects, which is up to explaining completely and exhaustively why that event happened as and when it did. This is as much as to say that no event, considered purely in itself and apart from any subjective consciousness of it, can be thought of as happening randomly. For if any such random event (that is, any event that 'just happens' and for which there is no full explanation) were possible, then *every* event would become objectively inexplicable – its explanation couldn't, in principle, explain why a (randomly) *different* event didn't happen instead. But a world thus wholly randomised couldn't be objective as over against any subject. The changes in the ice, for instance, taken as random, might just as well have been presented to an entity that itself changed continuously: no essential coherence would thus be lost, since there *is* no essential coherence in a sequence of might-as-well-be-random events. With randomness admitted, the idea of objective change, with the implied continuing subject as reference point, simply dissolves.

But then, if the set of reasons constituting a complete explanation of any event points us (as it must) to prior events that are together sufficient for the event in question, that event itself will have been a necessary occurrence – it will *have* to have happened – given those prior conditions. Given the full, third-personally available history of its changing states, the ice-cube not only is but *must be* in the particular state that it has reached at any particular moment of the melting process. This is the set of implications of objectivity (and so by extension of how the unchanging subject underpins our perceptual access to objects) that has long been labelled the Principle of Sufficient Reason.[6]

We have arrived here at the way in which a world considered as objectively confronting a conscious subject must, just as such, be considered as fully *determinate*. Translated into the terms of a natural world confronting a human agent, that yields the hoary philosophical crux of causal determinism. The problem arises because human agents are embedded in and (since we are Blackburnian naturalists) integrally part of a deterministic natural world in which every event has a cause – a precipitating set of prior events under the rubric of sufficient reason. Given its sufficient causal history, that is, every event must happen as it does and could not have happened otherwise. But human actions too are events, natural transactions in a natural world, and we simply cannot accept the application of this universal determinism to ourselves – because it would mean that to have acted differently from how I did act in any situation, I should have to have been a *different person.*

Why? Imagine that I choose a piece of cake rather than an apple at the lunch counter, with no-one else forcing or in any way influencing me to do so, and knowing that I could perfectly well have decided off my own bat to eat more healthily and choose the apple instead. I might say here that it was *entirely down to me* whether or not I chose the cake. But this cannot mean just that the causal conditions, given which the act of taking the cake *had to happen,* were all internal to me as agent. (They presumably involved things such as the state of my memory and anticipation, and maybe of my digestive system, at the relevant moment, together with all the past history that had generated a sporadically malfunctioning capacity for self-discipline.) Its being down to me has to mean something more than this – something about a centre of spontaneous willing – in something like the way in which the 'I' referring to the subject of perception, memory or desire has to mean something more than any organic and changing entity that could be me.

To connect this up with the terms of the foregoing discussion: when I picked up the cake, everything that had ever been true up to that point, considered objectively, must have included the sufficient reason, the full and final explanation, of my doing so. Since the relevant truths were internal to my nature and history, that means that the four-dimensional bit of objective reality comprising *me* (what the seventeenth-century German rationalist philosopher Leibniz calls my 'complete individual notion' or concept[7]) would have to have been in at least some particulars different up to then, for me to have picked up the apple. The totality of my properties, of *everything* true of me over time, would have to have been different, thus constituting a totality non-identical with the one that actually obtained. Difference between entities, considered objectively, can only mean non-coincidence of their properties, and must ensue on such distinctness. (If identical twins were identical in literally *every* respect, they would just be one single person.[8]) That is why, given the totality of my properties that actually did obtain, I *couldn't* have picked up the apple: determinism is just the outcropping in the domain of action of the principle of sufficient reason. I should have needed to be, objectively considered, a *different person* to have picked the apple.

When I know that I could, for all that, have taken the apple, however, what I know is that the principle of sufficient reason doesn't *reach as far* as the *I* in question here. Within the scope of that principle, it couldn't be the case that everything sufficient for my taking the cake was in place, and yet I remained free not to take it – I could only be in that position, so long as a full set of conditions sufficient to ensure my doing so had not yet concurred. To remain free to choose on the very cusp of action, as we cannot help thinking ourselves to be, I have to think of my willing self as something subsisting beyond the domain of sufficient reason. Underlying this is the truth that my acting from *this* subjective focus cannot be conditioned or affected by my specific causal history, since it is part of the idea of *being a subject* that I should be acting from this same focus with *any* history. That is as much as to say that subjectively I *couldn't* have been a different (Leibnizian) person, because from that perspective any way I happened to have conditionally turned out would still have been *me*.

Our need for this kind of subjective continuity across possible actions, I take it, is undeniable. If it weren't the case, regret (for instance) wouldn't make sense. Regretting that I didn't make some particular significant choice is thinking how *my* life, the life being lived now from *this* focus, could have been better – it is not envying the lot of some distinct but fairly similar person in a different possible but non-actual world. And similarly for related notions of credit and responsibility: I can only take credit, should I feel so inclined, for being the kind of person who could not (or at least, not readily) go out and commit rape or murder, if it is *not* the case that to do that kind of thing I should have to have been 'a different person'. I can no more intelligibly take credit for refraining from, than blame for committing, *a different person's* bad actions.[9]

And indeed, regret, shame and similar emotions, along with their interpersonal variants such as resentment and gratitude, *don't* really make sense on the objective view.[10] If to have done or not done X would have meant being a (Leibnizian) different person, and I am always exactly and only the person I am, having developed into that person in ways for each of which there is always in turn fully sufficient reason – then there would be simply no point in repining over anything. If it's no use crying over spilt milk, crying over milk which was always inescapably going to be spilt would be absolutely senseless.

Yet we know that regret and (underlyingly) responsibility for our actions are real. They can only be so, however, if the self that wills is subjectively outside the scope of the principle of sufficient reason – and that means, outside the objectively considered, third-personally rational structure of the world.

Self and environment

We have so far been considering this issue from inside consciousness, as it were – by exploring our experience of consciousness, and trying to get at the sense of a perceiving and willing self that lies at the core of it. But we can just as well look at the same issue from outside, by exploring the implications of our being organisms of the kind we are.

To think of myself as such an organism is to envisage something with a certain physiology, at a certain stage of its natural development through a characteristically human lifespan. With that goes a certain set of emotional and attitudinal dispositions, again at a particular stage of their natural development, resulting from a combination of organic endowment and interactive history. The organic entity that I am, that is, reflects a unique specific development through life-interaction with its environment and with other human beings.

This developmental history, too, is always ongoing – I am always moving in a complex, ever-shifting pattern of physical, personal, social and institutional relations, to which I am always both responding and contributing in a multiplicity of ways. As an organism, that is, I am inherently something that belongs in an environment. This is obviously not just a natural environment, as for instance a tiger belongs in its jungle, with certain kinds of prey living in its vicinity and so on. In the human case we are talking about an institutionally and linguistically shaped environment comprising people and their organised activities as well as places and things, one that offers a very wide variety of cultural as well as natural affordances. Correspondingly, belonging in it is a much more generalised and flexible business than in the case of the tiger, though not infinitely so. My identity is realised – that is, both formed and maintained – in ongoing exchange with this environment, and human identities can vary enormously more than tiger identities. But for all that, it remains true that any organism is not just what-it-is, but what-and-where-it-is – this is, after all, the fundamental ecological insight that underlies environmental thinking, when this is applied to human affairs (as human or political ecology).

It is a vital corollary of this insight that any organism removed from its natural environment has had a key part of its identity stripped from it, even if it can survive physically. (This recognition, of course, underlies the move from cages to would-be 'natural' surroundings in zoos.) *Belonging in* its environment is not a formally logical relation – it is not logically necessary that this particular tiger be in its jungle, since it doesn't stop being a tiger if you capture and extract it, and a sign reading 'Tiger' affixed to its cage in an old-fashioned zoo wouldn't therefore be making a false claim. But while this will be the case for any individual tiger, when we say 'the tiger is a jungle-dwelling creature' the connection we are making does seem in a sense to be a necessary one – we are recognising that it is part of being a *characteristic* tiger to be jungle-dwelling. Hence the relation is not a merely contingent one for any particular specimen either: *this* tiger doesn't just *happen* to be in the jungle, such that it could still be completely itself anywhere else. The relation is, precisely, *eco*-logical: the entity cannot be detached from its environment without radical loss, because *its being* is a process of interaction with its characteristic environment, and not something given independently.

And now the deep problem with thinking of myself as an organic entity can be put this way: I cannot think of *myself* as an organism integrated with its human context of significant interactivity, in that essentially ecological way – although I have strong independent theoretical reason, apart from my self-experience (the actual phenomenology of my being), for believing that this is what I am.

I can of course think of JF as such an organism, in a third-personal version of what I was doing a couple of paragraphs back. I can envisage *him* (as I can envisage anyone else) set in, and realising himself interactively through, a set pattern of natural and social relationships. But there is always going to be one more fact about this particular organism than the totality of facts that make such third-personal truths true, and that is the fact that this organism is *me*. Acknowledging that fact requires a radically *first-personal* thought, true not from each and every perspective, but only from mine. Such thoughts are made true by a particular relation between the subject and what is thought of. What that relation is, a famously vivid philosophical example brings out.[11] Imagine following a trail of sugar round the aisles in the supermarket, a trail obviously left by someone with a leaking sugar-packet in his trolley, trying to catch him up to warn him – and then realising that the person whom I am pursuing is actually myself. In one sense, thinking of 'the person with the leaking sugar-packet' is indeed to be thinking (third-personally) of myself throughout my pursuit, since, though still unbeknown to me, I *am* that person. But in another sense, and the important one here, I am evidently not 'thinking of myself' under that description until I have my flash of revelation and start thinking of the person with the leaking packet *as me*.

But if I do try to think of *myself* subjectively in this way, that is *me-centredly* or first-personally, as an organism in an environment, then I will find (as we have already seen in the previous section) that I am inescapably thinking of a self that is only contingently to be found in any particular set of relations, including its supposedly self-constitutive environmental relations. I think of myself, we might say, in such a way that I should *still be me* if I was permanently marooned on the Moon.

It follows that I cannot really think of *myself* as fully an organism-belonging-in-an-environment. I cannot do so, because to think third-personally, that JF is an organism so situated, involves leaving out the essentially indexical or first-personal fact that *this organism is me*, and so is not yet to be thinking of myself, while to do so first-personally involves thinking of something *as me* that is only contingently related to any environmental situation in which it stands. When I am thinking of myself, my being such an environmentally situated organism is, we might say, *essentially occluded*.

The self as constructed Ego-model?

In considering in turn how we appear to ourselves in perception, memory, desire and choice, we have found each of these functions subtending a centred self, something that we might label an *ego*, which has to be thought of as standing outside and over against the world that it cognises. We have found this ego, by the same token, to be something that seems not to stand in the same radically ecological relation to its surroundings as do all other creatures.

But to have *really* found this would surely be incompatible with our being naturalists in the sense indicated by Blackburn, to which I referred in the first section. It certainly does seem to involve an 'unexplained appeal', if not to anything supernatural. The problem here, clearly, is that we simply cannot understand anything natural as unchanging or undetermined, so anything unchanging and undetermined must be at least *a*-natural: not necessarily *super*natural, but outside nature.

Yet we are all basically naturalists of that Blackburnian kind now – which means that there is nowhere 'outside nature' to be. So what naturalistic sense can we make of this persistent and seemingly inescapable idea that there must be something like this at the core of ourselves?

An approach that has lately been popular among materialistically inclined philosophers of mind is the one glimpsed earlier as put forward by Daniel Dennett, who has long been the doyen of this kind of account. Basically it is to say: humans, as nothing more than the natural creatures they are, are *constituted as* 'selves' by (essentially) the reflexivity of their mode of consciousness. As a primate species they have had the brain size, the dispositional armoury, the linguistic potential and the environmental luck to evolve a form of consciousness that can include awareness not just of external or internal stimuli, but of its own processes of awareness of these stimuli. Human consciousness has evolved, especially through the development of language, an ability to turn attention upon its own operations of attending. This unique life-form, awareness of awareness (unique, that is, terrestrially and as far as we know), operates in a way that must naturally generate the impression (for itself) of *centredness*. For awareness of anything – that is, any form of consciousness – must always be from a perspective, and with awareness of awareness the standpoint for the perspective is evidently moving 'inwards'. That is just the constitutive intuition of a *centre* of awareness, a notional central point where we take the 'self' which is the bearer of this form of consciousness to reside. But actually, this entity is a construction or projection, an evolutionarily configured working model used by the human brain to organise the complex, temporally structured patterns of attention, to the rest of the body and the surrounding environment, of which it is capable. In the same way, the 'constructed self' is constructed as a causal force in what we experience as willed choice and action[12] – although underlyingly, the same unconscious psychological and physical events are impelling both the action and this 'construction'.

The idea of *self* does indeed seem to have a natural home in reflexivity – that is, the idea of an action that is directed upon the agent of that action. This is a perfectly ordinary idea with plenty of ordinary examples. A creature can, for instance, for some reason bite or claw at *its own* flank, and then we will quite naturally say that it is wounding *itself*. It can wound itself unconsciously and without intention, if (say) it bumps hard up against some sharp object when in flight from a predator. The core idea in either case is that what *is* wounded *does* the wounding, or otherwise causes it. But this simply will not do as a route to explaining consciousness, because the 'itself' of this natural reflexivity of action, and that of consciousness, are fundamentally different. We cannot just say, by parity of expression, that in reflexive consciousness 'what is grasped is what does the grasping'. There is an immediate oddness here, as in the notion of a hand trying to take hold of itself. That particular oddness might seem down only to the metaphor of *grasping* – but 'what is understood is what does the understanding', while more abstract, is no better. For what is understood is the *object* of understanding, and what does the understanding is the *subject* of understanding, and the concept of *understanding* as such seems to require, ineliminably, both of these components, and also that they be distinct. (An object of awareness must be for a subject, because the idea of subject-less awareness is a contradiction; equally, a subject can only be such vis-à-vis some object, because the idea of awareness without anything that is its object seems just as contradictory.) But then the subject, whatever inward, first-personal phenomena of attention it may be taking as its object, is not understanding *itself*, for it has to be understanding something distinct from itself to be understanding at all. As Kant, the greatest philosopher of the early modern era, puts it in a famous and still classic wrestle with this issue: 'It is ... evident that I cannot know as an object that which I must presuppose to know any object'.[13]

Presupposing the fundamental reality of the subjective self in order to explain this self in reductive terms (that is, as *not* fundamentally real), is in fact a way of begging the question into which naturalist accounts as such are always liable to fall. As an illustration of this, consider the account given by the contemporary German philosopher Thomas Metzinger. He argues that *taking itself to be* a 'self', something 'inside' looking out, is key to the way the human organism functions in its physical-and-social environment. This, however, is a kind of illusion, insofar as really (materially) there are only the organism availing itself of this model, its physical surroundings and its (socio-biologically describable) interactions with other organisms. Hence: 'Contrary to what most people believe, nobody has ever *been* or *had* a self.'[14] Rather, the organism *represents itself* to itself as a 'self' – as an Ego.

> First, our brains generate a world-simulation, so perfect that we do not recognise it as an image in our minds. Then, they generate an inner image of themselves as a whole. This image includes not only our body and our psychological states, but also our relationship to past and future, as well as to other conscious beings. The internal image of the person-as-a-whole is the phenomenal Ego, the 'I' or 'self' as it appears in conscious experience ... The phenomenal Ego is not some mysterious thing or little man inside the head, but the content of an inner image ... By placing the self-model within the world-model a centre is created. That centre is what we experience as ourselves, the Ego. It is the origin of what philosophers often call the first-person perspective.

Here one can see particularly clearly the question-begging move being made. The Ego or self is the content of an inner image, says Metzinger, and experiencing this image as central within our world-model is what gives rise to the first-person perspective. But an *image* requires someone to look at it. (Nothing could *be* an image in a world without observers.) So if the self is the content of this Ego-image ('the internal image of the person-as-a-whole'), what is this content imaged *to*? The only answer that the above passage suggests is: the brain, as agent of the biological organism. But that can *have* 'images', as can anything, only as *already essentially involving* a first-person perspective – that is, as already being a self of the kind that we are supposed to be having explained to us.

Attempts to explain the conscious self reductively, in terms of things that aren't selves, all characteristically beg the question in the way that this illustrates. That is, they rely on our basic self-understanding *as conscious*, in order to make the proposed reduction appear intelligible. But that understanding, as we have seen, essentially depends on the idea of something a-natural – and what, distinct from the natural, could an organism be turned into, just by the natural evolutionary development of reflexive consciousness? We are back with the problem, and the insistent appearance of radical dividedness, that materialist philosophers of mind were offering – brashly, as it now appears – to resolve. In order to understand our kind of consciousness, we seem to have to understand the core of ourselves, in perception or action, as in some radical way coming from outside the natural world. Yet we also know, as Darwinian naturalists, that anything we can be must be fundamentally natural.

The dark self

So what is going on here? How can division at this depth be accommodated in any sense that we can make of the human form of life?

The clue lies, I believe, in an idea briefly touched on at the end of the last section but one. In self-directed thought, our full being is necessarily occluded. The living ground of our being is necessarily *dark* to us – not psychologically dark, like the Freudian *id* that the ego suppresses from sight for its own security, but logically dark. There is something that is *my* being (something the totality of facts about which are not purely third-personal), but that is also not constituted by its own consciousness of itself from inside, as my subjective Ego-self is. The trouble is that as soon as I try to think of this situated being *as myself*, it becomes the 'I' that is essentially conscious-self-to-itself – and trying to think of what lies behind this is like trying to turn round fast enough to catch a glimpse of the back of your head. Yet all the while we know that, in the dark behind the 'I', there lies my radically situated, environmentally contextualised living being, informing my whole activity and experience through desire, impulsion, value and choice. As in cosmology *dark matter* is matter that is inferred to exist from gravitational effects on visible matter and background radiation, but is in itself unde-tectable by any instruments, so we may think of a dark self constituting each one of us, always inaccessible, yet needing to be assumed to explain our experience of what *is* accessible. The dark self is the whole being-in-context that we know we must each individually be, but that for the self attending to itself always eludes being grasped. The simplest way to put this also emphasises how close behind the ordinary and familiar this notion lies: the dark self is who I must be whenever I am not thinking of myself.

Lest this still be thought of as some kind of psychological point, we may put it in terms that are unambiguously metaphysical and epistemological. The dark self is the full being of the organic-conscious human entity as it exists, but as it cannot be *viewed* (in thought) from anywhere. It cannot be identified first-personally in the sense of becoming an object for the subject that it subtends, for the kind of reason we have already considered. It can't be viewed third-personally, because each third-personal perspective is one to which the crucial fact, necessarily available somewhere, that *I am* this object, cannot appear. Beyond both first- and third-personal perspectives this unique individuality could only be available to a 'view from nowhere', but there can be no such view – the idea of a view necessarily includes that of a viewpoint. That is why the whole self is *essentially* dark.

Yet as such, it isn't any magical or supernatural thing – it is an individual organic manifestation of the natural world as it is in itself, occluded under the necessary conditions of self-consciousness to which, in human beings, organic development has reached. Metzinger, for instance, is perfectly cor-rect to say that we as whole organic-conscious beings do not *see* the self, we see *with* it. Unfortunately, he fails to carry this insight all the way through. He supposes that with the aid of his 'phenomenal self-model' concept, aligned to the deliveries of neuroscience, we could come to see, and to analyse and describe third-personally, the whole organism's deploying itself as a self in reflexive consciousness. But if the Ego is this whole organism's instrument for attending to its own complex embedding in the world, then when through the Ego it attends to *itself attending*, it will still necessarily be encountering itself as a genuine a-natural subjective centre (in all the ways that we have been illustrating), and not as anything instrumental. The question is, why would it be wrong? This is, in other words, the question whether Metzinger and others who think like him are entitled to call the Ego-self an *illusion*. And by now it should be clear that the answer is no.

Originative a-natural subjectivity is fundamental to the first-person perspective as we have seen, and this perspective is ineliminable and basic to our kind of conscious awareness. An illusion that we can't help having, and than which nothing could be more real, isn't an illusion.

The nineteenth-century German idealist philosopher Arthur Schopenhauer was perhaps the first person to see this matter clearly. In his treatise 'On the Basis of Morals' he writes:

> The cognition which we have of our own selves is in no way exhaustive and clear to its ultimate ground. Through the intuition which the brain carries out on the data of the senses, thus mediately, we recognise our own body as an object in space, and through inner sense, we recognise the continuous series of our striving and acts of will originating on the occasion of external motives, and, finally, we recognise the various movements of our own will, weaker or stronger, to which all inner feelings can be traced. That is all: for cognising itself is not cognised in turn. In contrast, the actual substrate of this whole appearance, our inner essence in itself, that which wills and cognises itself, is not accessible to us: we merely see the outward; the inner is dark ... for the greater, indeed, the essential part, we are unknown to ourselves and a riddle ....[15]

And yet we know that the conscious self, our way of experiencing from within our organic being in the world, is and must be a natural product, fully unified with our naturally given embodiment as an evolved primate species. It should now be plain that the necessary darkness of this unity to understanding, the darkness for us of what must nevertheless be our real selves (organism and consciousness in combination) is what lies at the root of our inability to comprehend our own finitude, and thus of the metaphysical fear of oncoming death discussed in Chapter 3 – and thus, in turn, of our recoil from this fear into a pathological progressivism. And if this unavoidably long chapter has worked at all as intended, it should also be clear that this darkness of our whole, finite living being to itself is built into the very structure of being a human self.

To recall the pivotal questions identified at the start of this chapter, therefore, but now in sharper focus: How can radical inner dividedness of this order *happen* to a natural being? – and what, if anything, could we possibly hope to do about it?

Notes

1 Set out classically in Descartes' *Meditations on First Philosophy* (many editions).
2 Blackburn (1998), pp. 48–9.
3 Lawrence (1961), pp. 29–30.
4 'Metamorphosis', in Kafka (1992).
5 Dennett (2013), p. 334 – the following quotation is from p. 336.
6 The classic account is that given in Schopenhauer (1813/1974).
7 See the selection from his 'Discourse on Metaphysics' (1686) in Leibniz. (1686/1934).
8 Compare *Tragically, I was an Only Twin*, the title of the Complete Peter Cook (London: Arrow Books, 2003).
9 For the same reason, 'the person I am' cannot be my established character, habits and embodiment, despite the commonplace usage in which *person* really means *personality*. My 'established' character can change, for better or worse, and I can be held responsible for such change – certainly in character and habits, and often in embodiment – which could not be so unless 'I' stood for something apart from any of these embodied character-phases.
10 See P.F. Strawson's classic essay 'Freedom and Resentment' in Strawson (1974).
11 See Perry (1979).
12 See Wegner (2002).
13 Kant (1781/1999), p. A402.
14 Metzinger (2009), p. 1. The longer quotation following is from p. 7.
15 Schopenhauer (1841/2009), p. 264.

6

NATURAL RESPONSIBILITY

Someone might respond to the claims concluding the previous chapter: if *that* is its basis, doesn't all this talk of radical dividedness rather over-dramatise things? What force can the argument that our whole organic-conscious selves must be dark to our understanding really have, when we seem perfectly able to make sense of our being confronted with such an occlusion? We can surely grasp the way in which, as the bearers of a specifically reflexive form of consciousness, we are naturally liable to be so confronted – the evidence for our ability to make sense of that, being surely that we have just spent a chapter doing so. Why then should our inability to think our own finitude first-personally be taken to have such drastic implications, when we seem able to get a respectable handle in general terms on the nature of that inability?

That objection would miss the point. The kind of understanding that matters here is not that which we can have of an outline sketch of general relationships, but our urgent need to make *lived* sense of our own conditions of being. I certainly do know, on general evolutionary and ecological grounds, that I must be an organic-conscious unity, insofar as I know that I am human and that all members of the species *Homo sapiens*, must be such beings, and must be unified as such in order to survive. But that is very different from being able to *think of myself as* such a unity. To recap: to know any set of facts third-personally cannot be all there is to recognising myself, because the fact that something is myself is an ineliminably first-personal fact. But to know myself first-personally is to know myself as something that must, as subjectively conscious, stand over against the whole objective world that is knowable third-personally. I can know in neither mode alone something that seems to require being known in both modes, and yet the modes insist on themselves as alternatives. (If you still doubt that, try thinking of something as both 'it' and 'I' simultaneously; this is actually as impossible as lending yourself money, and for essentially the same reasons.)

Here is the crux that contemporary philosophy of mind has identified, and continues to struggle with, as the 'hard problem' of consciousness. That is the modern, neuroscientifically literate version of the classic mind–body problem. It focuses, as a good popular exposition of the issues recently put it, on 'the difficulty of understanding how physical processes in the brain can possibly give rise to subjective experience. After all, objects in the physical world and subjective experiences of them seem to be two radically different kinds of thing: so how can one give rise to another?'[1] But it is quite impossible to corral this worry as a 'merely philosophical' problem (whatever one of those might be). Taking it seriously means that in immediate personal terms, there can readily arise what we might call the *life*-problem of consciousness: a kind of unity is necessarily unknowable to me, which must be fundamental to my *whole life*. How could one over-dramatise that plight? How could its implications for making sense of ourselves be anything but drastic? How could it not cut so deep as to subvert our hold on both death and (therefore) life? – and how could the effects of that on our actions and natural relationships not also be profoundly subversive?

We shall return to these questions. Meanwhile, we need to reflect further on the darkness of the whole self.

The self-darkened self

Although we cannot give a naturalistic characterisation of our own inhabited self in its fullness, we can, as already noted, perfectly well sketch a naturalistic explanatory story of how the internal architecture of human being must have arisen. Evidently, awareness of ourselves as centred egos is the specific development into human form of the way in which all life is naturally centred as such.

To make this kind of centredness clear, consider by contrast a stone. Things certainly happen to it – weathering (as by frost), being shifted by a landslip, being submerged in a flood or by the change of a river's course, becoming algae-coated underwater ... – and through all this it exists as a referent ('*What* got shifted ...?'), and is in that sense central to a possible narrative. (We used, at least in my day, to write compositions on 'the history of a penny' in junior school, and could equally have done so for a stone.) But the stone is, in an obvious sense, *uninvolved* in what happens to it. Again, what happens to it may be affected by what it is like (whether it is granite or limestone, for instance, will determine what happens to it under particular pressures and stresses), but this is not interactivity in the same key sense. What that key sense is, we recognise immediately as soon as we contrast the stone with any kind of living thing. A living thing *responds*, however primitively, to what happens to it – it is different in differing environments. (The stone is itself wherever it is, and so lacks 'an environment'.) Any living thing is constituted by a reciprocal process of interchange with its environment, through which it is always ongoingly establishing and maintaining itself. This is *autopoiesis*, the process whereby, from the cell upwards, living systems organise themselves as persisting bounded structures permanently interacting across those boundaries with a sustaining environment.

Once this is in place, it is a perfectly natural progression to the infinitely more complex life-centredness of a human self. The autopoietic centredness of life develops into conscious awareness. (Note the root of *a-wareness* in *wariness*: an organism is better able to cope with threats from its environment if it somehow represents that environment to itself – that is, to the extent that it responds to some sort of representation, which initiates it into the use of memory and anticipation, rather than simply responding directly to immediate impingements from the environment.) Any organism as such behaves autopoietically, but animals constitute an evolutionary advance on plants in virtue of a more complex environmental responsiveness, and higher animals by including awareness of their environment within the business of responding to it. This happens with hugely differing degrees of sophistication, obviously, depending on the life-form in question – but consciousness of any kind *must* stabilise the subject in relation to its objects in order to function. It must thus 'subjectify' the perceiver as well as 'objectifying' what is attended to. (We can see this easily enough in our pets – Tilly the retriever registers the changes in position of Fred the Furry Pheasant as he is tossed across the room, in relation to her own unchanging perceptual focus – and thus catches him.) Human consciousness adds awareness of the subject *as* thus stabilised and 'subjectified', and of that which is attended to *as* objectified. At the highest animal levels the primitive forms of self-consciousness are already apparent, but it is only with the much more complex brains and cognitive capacities of human beings that we reach animals possessing *awareness of themselves as aware* of their environment (now including their 'internal environment'), that is, the fully reflexive kind of awareness in which resides the self-recognised and self-possessed living centredness of the human self.

Equally evidently, the contribution of language to these developments must have been crucial. Before language could get its start, the sense of conscious selfhood must already have been strong enough to organise the inherently phenomenal character of experience into communicability. 'Wolf!' (or whatever articulated grunt initially served that function) must from the beginning have carried the sense of: '*I* am aware, from over here, of a wolf, and so if you paid attention could *you* be from over there.' Language not only enables us to meet in meaning but presupposes different perceiving foci capable of profiting from so meeting. It is likely enough that self-awareness as 'me' only crystallised out from 'awareness of awareness' under the survival and group-organisational pressures that must have elicited such proto-linguistic exchanges from creatures with an apt vocalising apparatus. This 'me'-ness would then carry over as the linguistic-conceptual nexus, *first-personality*, within which a human being would relate himself to the experiences that he undergoes. It would, that is, come to register its life-centredness to itself in a way now essentially structured by logic.

Subject and object, in other words, are not detachable phenomenologically, in the basic mode of life-centred awareness. As we have already noted, an object has to be object-for-a-subject, but a subject without objects is also inconceivable. Object and subject of consciousness as such stand in permanently reciprocal relations. But logical subject and predicate must be distinct: 'I *see a wolf*' only predicates if I as subject am distinct from the wolf-seeing – 'S is F' could communicate nothing if, were S not F, it wouldn't be S either. Once communication comes in, and with it criteria of truth and falsity, subject and predicate must fall to opposite sides of a logical – thus an unbridgeable – gap. When self-awareness is scaffolded into first-personality by language, it must differentiate itself radically from its objects. While awareness simply as such might have remained in subject–object symbiosis – organic consciousness as subject-*in*-environment, not self over *against* environment – a reflexively conscious selfhood that we have to experience as radically a-natural, with all the associated problems of self-understanding that we have been exploring in the previous chapter, seems to be the evolutionary price that human beings had to pay for language.

Living from the dark self

We might then well ask: how (on Earth) were we able to afford this price for so long?

It would follow from the foregoing sketch that human beings have lived with the inability to grasp in explicit self-understanding the full organic-conscious completeness of their being, virtually since the dawn of their articulacy – and certainly since they started thinking about it. But we need not take an unduly roseate view of human history to observe that they seem, nevertheless, to have coped. A sense of radical inner dividedness and the accompanying metaphysical fear of death and finitude, that is, must – if the sketch is even roughly right – have been present as real threats to human being from the beginning. Evidently too they must have become more than just potentialities from time to time, in spiritual crises of various kinds from cults of human sacrifice through the mediaeval Dance of Death to the obsessions of Calvinism, for instance. But humanity has clearly not been disabled in the longer run by these potentialities and crises, for the business of building up the human world, with all its complex cultural forms of life-meaning and life-satisfaction. Simply, generations upon generations have succeeded in living and dying as full human beings, however materially and technologically impoverished the vast majority of those lives will have been. Had that not been so, our species would long since have perished in massive existential breakdown. It is only comparatively latterly, according to the account given in Chapters 2 and 3, that we have begun to incur the pathologies of progressivism that now threaten such an outcome: a process that we can now think of, in this context, as the escalating metaphysical and then (consequently, via progressivism) environmental debt into which we are locked, once we can no longer, for whatever reason, *afford* the life-price of human consciousness. But where have we previously, for most of history and prehistory, found the necessary resources to afford it?

Schopenhauer, whom I have quoted before, gives a strong hint in a passage where he is discussing what I have already suggested as the test case for radical self-division – the confrontation of the individual with oncoming death which we considered in Chapter 3. In contrast to any metaphysical fear of unintelligibility, which would drain life of its point, he points to the certainty that we can all at least glimpse deep in ourselves, that actually *life is its own point* and death is always going to be beside that point. Without some such non-cognitive blood-certainty, it is evident that the considerations explored in that chapter, available at some level to almost anyone who thinks with any persistence about the issues, would either generate some hysterical form of escapism in essentially the way I have characterised as progressivism, or would simply paralyse action – and equally evidently, for most of past human life, they have done neither of those things. Nature, the mere self-warranted ongoingness of life-activity, must for all this time have been able to insist on itself in face of any unintelligibility afflicting subjective consciousness considering its own mortality.

Schopenhauer indeed writes so perspicuously about this point that it is worth giving it, again, in (the standard translation of) his own words:

> The animal ... lives fearlessly and heedlessly in the presence of anni-
> hilation, supported by the consciousness of being nature herself and
> as imperishable as she. Man alone carries about with him in abstract
> concepts the certainty of his own death, yet this can frighten him only
> rarely and at particular moments, when some occasion calls it up to his
> imagination. Against the mighty voice of nature reflection can do little.
> In man, as in the animal that does not think, there prevails as a lasting
> state of mind the certainty, springing from innermost consciousness,
> that he is nature, the world itself. By virtue of this, no-one is notice-
> ably disturbed by the thought of certain and never-distant death, but
> everyone lives on as though he is bound to live for ever.[2]

Man, that is, like all the other creatures, is fundamentally 'nature herself',
on which death has no hold.

But if humans have been able to *put* their own deaths beside the point in
that way, still, as creatures who live cognitively as well as intuitively, they
have also needed to be able to make habitable sense of how and why they
can achieve this. In this context it is very relevant that Schopenhauer goes
on, from the passage just quoted, to note that:

> The same thing can also explain why at all times and among all peo-
> ples dogmas of some kind, dealing with the individual's continued
> existence after death, exist and are highly esteemed, although the
> proofs in support of them must always be extremely inadequate,
> whereas those which support the contrary are bound to be powerful
> and numerous.

So we might be led by this to suppose that, if we can no longer make habitable sense of the life-intuition that death is beside the point, that is a matter
mainly of such religious dogmas' having now effectively lapsed, with the
associated supportive practices and ceremonies falling into desuetude – at
least in the advanced West, where it most matters. (The West is where
progressivism is most pathologically embedded, even though it is through
the imitation of the West by the East that it now begins to operate most
destructively.)

This, however, is too simple as it stands. It has, after all, always been
deeply problematic how the prospect of simply continuing to live after
death could provide *any* sort of answer to questions about life's point. If
conscious experience just as such lacks point through inherently lacking
completability, why should further such experience, in however removed
or transfigured a form, be supposed to supply it? Responses to this query

often invoke the ethical dimension so stressed by moral philosophers such as Kant and Sidgwick: the thought that, since death's being the end seems to make the life of self-discipline and moral striving into a kind of fraud, we must envisage a necessary beyond-life as the domain of reward or punishment, that is, of consequences commensurate with our differential moral worth. But it is far from clear that this, either, makes the required kind of sense. We cannot really imagine *endless reward* for a moral agency that would then have ceased – the idea of reward works only within the stream of action, either as an incentive or as a restoration of balance so that we can go on together. (You find and return my wallet – I am in your debt, I reward you: now we are square.) Endless rewarding makes no more sense than infinite repayment: if a debt is finite, there will come a point at which I have finished repaying it; if it is infinite, it is not clear why I would even start. If post-mortem rewarding is at some point to stop, however, so that we must then contemplate, presumably, a few aeons cheered by fading recollections of it and then mere blank eternity, that reopens the issue of meaninglessness with a vengeance. And the same holds even more clearly for punishment: *endless* punishment for a period of moral agency that is over cannot constitute anything but intolerable injustice. But if it stops when I have paid my debt and am purged, what then? These questions insist on themselves if we take moral agency to conclude with death. But, of course, if moral agency and fallibility are supposed *not* to be over, but to continue after death, then the whole problem simply recurs.

The religious dogmas and associated practices to which Schopenhauer adverts, that is, could never really have helped with our metaphysical fear of meaninglessness in face of death, even when they were in full force, if they had been taken (whatever they explicitly *said*) as really offering to console us with *im*mortality – with some version of life's somehow going on for ever. They can only have been as potent as they once obviously were in reconciling human beings to their situation, by standing as a living index of our *a*-mortality: that is, of oncoming death's irrelevance to what we essentially are in the only life that we actually have. And they could have that kind of effect only because they were actively borne up, through most of their history, within the flow of the utterly non-dogmatic intuitive consciousness to which Schopenhauer also explicitly adverts: the awareness of our living selves *as nature*.

The sense of living nature as its own point goes very deep. And here we can begin to see what we have now put ourselves in a position to be able to say, in answer to our second pivotal question. *Why is wildness a necessity?* What have we really lost by losing living contact with it in the modern era, and how could re-attunement to it help us?

Wildness characterises whatever lives or acts from the whole of itself – whatever is steered in action by its whole self-in-environment, and is thus always wholly itself in living and acting. Wildness must therefore be clearly distinguished from *wilderness*, though that will obviously be where people who recognise a need for it will now tend to go looking for it. Wildness is not a location, nor an ecological or ethological classification, but a state of being. An organism is *tamed*, by contrast, to a greater or lesser extent, if it is steered or constrained in action by something other than its own whole-ness – either from outside itself by another will, like the tiger in its cage, or from inside, by an impoverished or conflicted self-recognition, as humans are uniquely in a position to do to themselves.

Humans, that is, have always the liability to make for themselves structures of being and patterns of behaviour which alienate their own wholeness out of their active grasp. They have always the potential to act only from their inevitable subjective *me*-centredness confronting a law-governed realm of objectivity – to take themselves, individually, as what D.H. Lawrence dubbed 'little mortal Absolutes',[3] seeking gratification in an externalised world from which they are essentially divided and that is therefore taken as essentially instrumental to their wills. They are always disposed, in the terms that we have been developing, to lose their living sense of their whole dark selves.

In the next section we will glance (there is space for no more) at some central manifestations of human being – in creativity, in exercising the vir-tues of open exploratory learning, in commitment to ethical and prudential claims that confront us with intrinsic value, and in love for other people – where only acknowledgement of the inaccessible presence of the dark self can make sense of what we are really doing, and correspondingly where a loss of living awareness of that presence tends to make us unintelligible to ourselves, and thereby threatens us with radical damage. Let us stay for the moment, however, with the central case of our confronted mortality. We can now recognise the possibility that, if we can somehow retain awareness of our underlying spontaneous wildness, we can live in the application to ourselves of the truth that what is wild is fully present and complete in itself at every moment. It is so in the dynamic way that is essentially characteris-tic of life – a plant or animal is always *achieving itself*, autopoietically, from moment to moment as a stone or a crystal is not. But since anything wild is always self-achieved in this way at every moment in its life, its death at this or that moment is also always beside its living point. Equally, it can never be incomplete because it is always, underlyingly, wholly itself, wholly achieved, in a way to which not only its lifespan but its life-conditions are irrelevant – the thousand-year-old sequoia or the hunting hawk no more and no less than the snowdrop trodden flat before it can open, or the half-grown rabbit stooped upon. Death simply happens when it happens and doesn't matter.

It is because human beings have for most of their history lived in ways that have kept them intimately in touch with that sense of the naturally wild, and have preserved a living sense of themselves as attuned and akin to it, that religions have been able to remind us of our a-mortality while talking (officially) about life-after-death. As Blake says, with his characteristic luminous simplicity:

> The wild deer, wand'ring here & there
> Keeps the Human Soul from Care.[4]

The deep understanding of life's being its own point going with this constant, familiar and day-to-day drifting of wildness through our lives has hitherto wordlessly recuperated our tendencies to both explicit dogma and supernaturalism, retrieving them to supply our absolute need for life-structures in which to put oncoming death aside. 'Wherever your life ends, it is all there', wrote the sixteenth-century French essayist Montaigne, perfectly capturing the essential recognition here.[5] Imagery of the forgiving creator, before whom at the last we stand whole, transitions easily enough from this insight towards religious claims, practices and institutions; but the roots of the insight itself are evidently in that awareness of our fundamental kinship with all of nature in spontaneous wildness which Blake's image aphoristically expresses, and on which Schopenhauer is insisting.

Such awareness is what I intend by the term *natural responsibility*. Its deep connection with our environmental situation should be apparent. Fundamentally, it is the betrayal of such responsibility, for what have seemed to us compelling reasons of human development, that has led us into environmental tragedy. But what we are failing in, here, is not to be confused with the economistic concept of responsibility central to the sustainable development paradigm: the responsibility of presently existing people either to our own later selves or to future generations, for the state of the planet's ecological resources as we hand them on. That is the superficially pragmatic but actually quite inoperable notion of responsible action that in *The Sustainability Mirage* I labelled 'shadow stewardship'.[6] Natural responsibility, by contrast, is the kind of inner self-responsibility, to the whole of our humanness, that we acknowledge when we try to act from our inherent capacity to flourish as fully the kind of creatures we are. But it is *natural* responsibility because living from our wholeness in this way depends on our being able to identify and answer to that in our nature which needs its full context in the wider natural world to be grasped and understood.

Inhabiting paradox

Alertness so fostered to the permanent background presence of the dark self is needed to make sense not just of our mortality, but also of some very central elements in our ongoing lives.

To illustrate this, consider in the first place *creativity* – meaning thereby nothing arcane or too precious, but the ordinary capacity that we all share to a greater or lesser extent for realising ourselves in forms of symbolic action. The issues are clearest, perhaps, if we think about something that will be familiar to most people, the attempt to express oneself adequately in words – generally (since this is usually where one takes most care with such expression) in an extended piece of writing. Here a very common experience is that of working towards a rightness, the criteria for recognising which one can't yet specify. One knows that one will know *when* it comes right, but not *how* one will know. At that moment, what is seeking form through one's attempts to 'express oneself' will suddenly speak clearly.

But what is involved in this process of finding the right expression for a thought? What *guides* us here? Not the thought itself as conceived independently of its expression – that was the eighteenth-century model of 'clothing the idea in suitable words' (a model arising mainly from the habit, promoted by a classical education, of trying to improve on thoughts that had been expressed before). Genuine thought, however, does not in that way precede its verbal expression, but rather is only realised once expression has been found. ('How do I know what I think until I see what I say?', as E.M. Forster is supposed to have asked.[7]) But nor by the same token can the guidance that we recognise be merely our knowledge of the meanings of words, since that could only guide us in aligning words to precedent thoughts. And yet, what more can there be to the process of self-expression than my putting a thought into words of which I already know the meaning?

What these difficulties should alert us to is that rightness of expression (by a standard that isn't producible until something is judged right by it) is given to us, when it is, from somewhere intimate but inaccessible. In the tension between realising my experience in a shared vocabulary, and getting it right (as what I myself *really mean*), I acknowledge my responsibility towards a whole self that has meanings – recognitions, intentions and impulses – that are not *mine*, in the sense of belonging to my conscious ego. Language in fact *can* only mean because it is dark selves that are communicating. (If each language-user acted as though meanings were things at the disposal of the ego-centred conscious self, language could never have got started, because such meanings would be inescapably private.) Word-meanings bespeak the inaccessible whole being behind the 'I', which *means* out of its organic-conscious participation in the common pool of human reaction and possibility, and when words fall into their inexplicably right places, that whole being is expressing itself through the ego which cannot grasp or explicitly acknowledge it as such.

Again, consider *learning*: the ordinary unspectacular business, say, of acquiring a new skill. Think for instance about learning to drive, where the pragmatic paradox involved in mastering any skill (you learn by doing something you *can't* yet do) is particularly evident. You learn to coordinate hands, feet and eyes in putting the car into motion and controlling its starting speed, by 'letting it come', that is, by letting the consciously willing ego step back. As you feel for the right moment to declutch, let off the handbrake and slightly accelerate, the coordination of what you are doing is entrusted, always a bit riskily, to something with an incipient competence that you don't yet consciously command. We practise ourselves, as it were, in a poised submission to the emerging embodied know-how of which we are trying to gain possession.

And then, by extension, think about what learning involves in any complex area: learning how to understand an unfamiliar topic (a new subject area, a new job), or acquiring new life skills, the sort of self-development that goes on as one merely moves through life and encounters new kinds of situation and life-problem. In all these activities we are most effective as learners when we are able to act despite uncertainty and discover from what happens which path we are actually on. As La Rochefoucauld observes, 'we come quite fresh to the different stages of life, and in each of them we usually lack experience however old we are'.[8] But there are rules of thumb: make as few working assumptions as you can, find your way forward by trying things out to see what runs, be aware of reactions, be ready to shift your weight, rebalance, move responsively with what emerges. We learn on the job at any level, indeed, by acting (as far as we can) in ways that open up options and deliberately don't decide prematurely between them.

Correspondingly, the relevant dispositions and character traits here are virtues – life-strengths in the exercise of which we thrive. They might be summed up as the capacity not to shirk the emergent, not to be always accommodating the new to the familiar or domesticating potentially challenging experience under concepts with which we have grown comfortable. These virtues exhibit essentially the same kind of paradox as we have been noting in relation to creativity: they must be strengths for winning by surrender – for negotiating the permanent existential hazard of committing oneself in the direction of something one does not yet know how to recognise. (As Socrates says in the *Meno*: 'We shall be better, braver and more active men if we believe it right to look for what we don't know'.) Learning of this order exercises what Keats famously called *negative capability* – the capability to not be restricted by what we are while we find out what we can creatively become.[9] This is paradox, but habitable if we are livingly alert to the dark self. To learn anything important, I have to act on the intuitive recognition that my whole dark self can be trusted to find its way towards and through more than I can ever yet consciously say.

Third, consider *value*: the utterly commonplace feature of human life that we act because we judge that things matter. This feature, however commonplace, notoriously bristles with paradox as soon as we start to think at all closely about it. It is fundamentally *what I want* that accounts for what matters to me and thus motivates my action – but value motivates me, and yet is supposed to guide me in what I *should* want, what *should* matter to me. Honesty (for example) is not something that matters because I desire to be honest (because then I might just cease to desire it), but attachment to it motivates my not lying (when on moral grounds I don't) in a way that only a desire of mine seems able to do. And so on, familiarly enough to anyone who has ever had any dealings with the philosophical field known as *meta-ethics*.

But as we have seen, it is paradoxes such as these that show us aspects of what we fundamentally are, appealing to the presence of what lies behind the 'I'. In relation to value, this is a matter of putting us onto the deep structure of motivation. If I desire S, I am to that extent satisfied by getting S, and the desire is placated. But the impulse given me by my values cannot be placated in this self-focussed kind of way. (I can stop at the third glass and look forward to more tomorrow, but I can't honestly think: right, that's enough honesty for today.) Again, in regard to any one of my desires, I can think that I might be better rid of it ('I could really do without having this hopeless yearning for … '). But I cannot think anything like that in regard to any of my values. If I can seriously think that I could do without being quite so hung up on honesty, I will not be properly honest at all. A desire is a lack that is conscious of itself and of what would go to fill it. But value seems nothing like this – to value honesty or beauty seems not to be moved from any kind of ego-centred need. So we find ourselves having to recognise that value must come out of my whole being. What I really value, what I value intrinsically ('for itself'), is that to which I intuit that my whole self, from the dark behind my consciousness, is committed.

These are commitments that I can't explain as choices or contingencies of my ego-self, but can only *live out* as determinations of my whole being, expressing its whole interaction with others and with its surroundings, a pattern of what matters, which I as ego don't choose, but must live from within.

Finally, consider *love*. This, too, has to be an arena where we are intuitively responsive to the dark self if we are to make habitable human sense. Once again it is Philip Larkin who helps to bring this out most clearly by contrast, wryly identifying the difficulty we have with this emotion as one of being *selfish* enough

> to upset an existence
> Just for your own sake.

If one were only the logically private ego confronting a determined objective world, this *would* have to be the real story about the sort of impulsion that carries one towards another. After all:

> My life is for me.
> As well ignore gravity.[10]

When I *think of* myself, 'My life is for me' is a truth that does indeed seem as unignorable as that. 'My life' under the rubric of such thought is that of which I am the inevitable subjective centre, the point of reference for all immediately experienced perception, feeling and willing. A way of being in the world that has me thus as its centre is then one in which choosing my pain over my pleasure – choosing to 'come off worst' – simply doesn't make sense. But actually we know (and perhaps most clearly in the experience of love) that my life is not 'for me' in that way. My life is *for* those with whom I am bound up just as much as for myself – for my wife, for our daughter and sons, for my parents and for others whom I have loved, and for my students, and for the legacy of a great teacher, and for this writing, and even (however unfashionably) for England .... 'My life', at the full, is that of an integrated mind and body and locus of agency essentially involved with others and with its surroundings, a speaker of a given language at a particular time and place within the history of a particular culture. Nor is this just a more elaborate way of thinking of an organism in its widest setting – 'all my contingencies' – since such a socio-organic entity isn't *objectively bound* by anything. It is just set in certain kinds of relationship; the *for* in 'its life is for ... ' is just a functional term. But the way my life is *for* my offspring, for example, is that I am really bound by ties of love, loyalty, paternal duty and affection ... I cannot see these as just 'my contingencies' and genuinely recognise them.

As we saw in relation to value, there are things that stand firm for us in a way that is deeply paradoxical either from the perspective of 'My life is for me' or from that of a particular social organism. There are loyalties that we don't *have* (either as evaluable from the me–centre, or as functional contingencies), but which we *are* – allegiances knitted somewhere in the dark behind subjectivity and objectivity. Without loving relations based on such loyalties the human spirit corrodes, and we are driven to seek escape from ourselves in the kind of permanent restlessness for which progressiv-ism provides warrant.

Wildness and sense-making

In relation to all these vital ways of making sense of our lives, I have invoked awareness of or intuitive responsiveness to the dark self. Does that mean that our wholeness is not, after all, dark to us?

Matters are more complex here than that simple either/or would allow for. To put this back into the terms of our previous discussion: humans have historically been able to afford the evolutionary price of language and reflexively conscious awareness, because they have always hitherto been more or less able, if not to grasp in explicit description their being dark selves, nevertheless to *live from* these selves – which means, to live in tacit responsibility to their full organic-conscious unity of being. The dark self is the self as wild. It is the self always lying beyond what is available to our individual conscious self-understanding. This self-understanding is always threatening to tame our actions and thoughts to the will of the conscious Ego. But through attunement to a surrounding wild world with which even the most developed earlier civilisations were always in close and intimate contact, human beings have nevertheless been able to inhabit a working sense of their whole selves, and let that sense tell in action and thought, even while the full underlying unity of the conscious self with its embodi-ment has always (and necessarily) eluded explicit cognitive formulation.

The key word in that paragraph, however, is *hitherto*. What we must now recognise and confront is our increasing inability under current condi-tions to inhabit cognitive paradox from the intuited resources of our whole selves. Modern man's conviction of kinship with wildness, Schopenhauer's 'certainty, springing from innermost consciousness, that he is nature', is now very far from what it was. A sense of our life as attuned to that of wild nature does not easily coexist, to put it mildly, with forms of liv-ing both physically distant and practically alienated from the unfenced and unmanaged natural world, as are the vast majority of contemporary urban or megalopolitan lives. That these forms of living typically offer instead a riot of manufactured stimuli to the ego, the self-centre that can be gratified by its sensory, cultural and shopping opportunities, serves only to confirm and deepen the alienation. Nor can vacational exposure to the non-urban, even in increasingly exotic places – in effect, an elaborate extension of shopping – go any significant way towards restoring this lost relationship. We are now confronted at every turn with the same *made* world, the same exhibition of human command and control, extending right across the planet. The effect of this loss of wildness is a loss of capacity to make living sense of ourselves, over crucial ranges of human existence.

We have seen that failure most dramatically displayed in our recoil from oncoming death, our deep bafflement at our own mortality and finitude. But there are ever more visible signs of our culture's inability to make human sense in other areas – of the increasing dominance within contemporary civilisation of the assumption that 'my life is for me' expresses mere unchallengeable common sense. I have space here to do no more than gesture at some of the indicators of this plight. When, for instance, in the discussion of any current issue of value did you last observe its being taken for granted that people have binding moral obligations that they do not choose, or that they can commit themselves to things from which they cannot then morally withdraw, or be released even by the other people affected? Or when did you last read any published advice column whose assumptions and values in regard to our intimate 'relationships' were not all in the lethally corrosive spirit of 'My life is for me'? We are always, as we have seen, inherently threatened with such radical self-impoverishment, with each being reduced to an a-natural satisfaction-centre over against a determined and essentially instrumental world; but in modern conditions the potentiality becomes ever more and more of a reality. And the recoil from it into progressivism follows in the way that has already been sufficiently indicated.

Can we nevertheless escape the self-generating bars of this mind-cage? Can we still find ways of living in permanent tacit acknowledgement of our dark selves? Can we again bring to bear on the vital transactions of life the whole being, the full complex socially and ecologically situated, articulately conscious organism that we each are? How far do the full potentialities of our species life-form, that is, remain open to us to be realised? These are the questions on which any prospect for *hope* – the guiding theme, it should be recalled, of this part of the book – must depend.

As the American philosopher Martha Nussbaum remarks, 'All living is interpreting; all action requires seeing the world *as* something',[11] and this must be true for any conscious creature. But human beings, who have risen to reflexive consciousness, live in an ongoingly represented manifestation of the world, of which we can always become aware *as* represented. Thus human being is epistemically configured at its roots. What this makes plain is that if we are to flourish as beings with this form of life, we need to pursue sense-making out of the *whole* of ourselves. If we confine ourselves to acting from the ego, on the subjective–objective model of self and world, which finds the merely cognitive (explicable knowledge) so congenial to its underlying principle of sufficient reason, we will miss most of our potentialities for sense-making, and so for living fully as human beings. We will then find ourselves having to operate with a radically impoverished picture, and in particular, we will find uninhabitable the paradoxes by which we negotiate the central demands of life and of its boundedness by death.

Wild creatures that live automatically and unquestioningly from the whole of themselves make by nature the best sense of their life-worlds that is open to them. Making the best sense of our life-course that is open to us means living in the same way from a specifically human spontaneity of being. But this suggests a vision, held up before our minds for guidance, of how we should behave as whole, un-self-impoverished selves. And on the argument that I have been assembling in these two chapters, such a vision is necessarily denied us – because, as we have seen, as soon as we are thinking of ourselves in relation to our recognitions and options at all, we are inevitably living through and as the ego–centre subtended by reflexive consciousness.

An ideal of *human* wildness, that is, considered as any set of criteria for action and thought, must turn out to be essentially occluded from us, in exactly the same way as that in which the whole unified self is dark to us. Yet we need to steer by an ideal of wildness in order to make sense of our lives as those of integrated and finite selves.

That is why we have to take *non*-human wildness, the wildness of external living nature, as our guiding ideal. The lovely vehement self-insistence of untamed things offers us a standard of active wholeness, an index of intelligible spontaneity and a permanent enactment of flexible responsiveness to the emergent. We might say that it offers all this to us as a living symbol, as long as that is not taken to mean an emblem standing for something identifiable and accessible independently. The wildness of the non-human world, in our acknowledgement of it as pervading external nature, *is* for us the spontaneity that in the same movement of attention liberates us inwardly for the hope of making a habitable human sense of ourselves. Living as far as we can in and from that kind of attention is what I mean – and what I have needed these two last chapters, and really the entire book up to this point, to be able to mean – by the permanent, exacting demand of natural responsibility. And our vital life-need to rise to that demand is the deep truth of Muir's claim that 'wildness is a necessity'.

Nature and 'nature'

Putting it like that, however, brings into clear view a large issue that has peeped over the edge of the discussion from time to time hitherto, but which we can no longer postpone addressing directly.

To have got ourselves locked into progressivism, I have been arguing, we need in some sense to have lost contact with the wild in ourselves (that is, with our whole organic-conscious selves, which as such are dark to their own consciousness). That means, on the account of the previous section, to have lost contact with the *external* wild: to have gradually lost the ordinary daily experience of this external wildness, and so to have lost hold of an essential guide towards an understanding of our whole selves.

But how, it may be asked, can the external wild *guide* us towards anything, when its naturalness can only exist for us by virtue of conceptual work (of identification, categorisation and the drawing of distinctions) that we must have *already* done? Wouldn't this be like letting yourself be 'guided' by a mind-reading Satnav that spoke back to you the results of simply following your nose?

Here we have the problem of the *constructedness* of nature (or, of 'nature') which has dogged discussion of these issues for a long time. It has often emerged latterly as an ideological conflict, between, on the one hand, deep-green radical ecologists who want to point us back to the intrinsic value of a nature standing over against humanity, as a criterion for setting civilisation to rights, and on the other hand those of a more Marxian or at any rate sociological perspective who see in any natural context, and thus across human–natural relations, the expression of power, class and (more recently) gender relations among human beings. A good account of that tension, and of why both sides of it are inescapable, is given by Kate Soper in her excellent book *What is Nature?*. She makes the hardly disputable point that

> while it is true that much of what we refer to as 'natural' is a 'cultural construct' in the sense that it has acquired its form as a consequence of human activity, that activity does not 'construct' the powers and processes upon which it is dependent for its operation … it is not the discourse of 'global warming' or 'industrial pollution' that has created the conditions of which it speaks.

And yet, as she also notes:

> There are many reasons to be sceptical of those discourses [of 'nature']
> that see no need for the inverted commas ... the discourse of 'nature'
> has served mystifying and oppressive ends, whether in legitimating
> divisions of class, race and gender ... or in promoting fictitious con-
> ceptions of natural and tribal identity that have been all too destructive
> in their actual effect.

She adds, however, in a nice demonstration of the permanent tension
at work here, that 'none of these points about the constructed nature of
"nature" ... can be registered without at least implicitly invoking ... extra-
discursive reality'.[12]

As this last remark should make plain, what we have here is not fundamentally a political or ideological tension, but a conceptual or metaphysical
one. It is also one that appears to threaten trouble for my argument, insofar
as that depends on the idea of our having (and now increasingly losing) a
guiding sense of contact with a nature external to the humanly organised
and managed world.

For an example of the kind of trouble it threatens: one might think that
a clearly identifiable stage in our move from a civilisation in daily touch
with its pre-urban roots in the natural to one surrounded overwhelmingly
by machinery and the artefactual was the supercession of the horse by the
internal combustion engine as the major motive power for urban transpor-
tation. It is certainly a matter of empirical observation that we no longer
have these large and impressive beasts trotting along our ordinary streets
with carriages or hansom cabs in tow. Here if anywhere, surely, we can put
a finger on a simple way in which contemporary life has less routine contact
with the natural (and thus, at a remove, with the wildness of the originally
natural) than it had in earlier phases.

But then – why is a horse more *natural* than a car? This question takes us
back to a distinction formulated most clearly by John Stuart Mill.[13]

In one sense of the term, they are obviously as natural as each other.
Both are produced by the powers and forces of the physical world, working in accordance with the laws of nature – along different causal tracks,
obviously, but nevertheless both outcomes of the causality that runs
through nature from end to end. But we also use 'natural' in another
sense, to mean (as Mill puts it) 'what takes place without ... the voluntary
and intentional agency of man'. In that sense, clearly, the horse is natural as the car is not – its existence and species nature are not (apart from
a little tinkering around the edges by breeders) the intentional work of

human beings, whereas the composition and functional organisation of the car are wholly due to such work. If we are reading 'natural' in that latter sense, however, what happens to the idea of 'guidance from the natural' towards understanding ourselves? For 'everything that doesn't happen through human intentional agency' is not a category that simply crops up in the world apart from human concerns and self-understanding (as, by contrast, birds or igneous rocks, for instance, seem to do). Without a prior and fundamental understanding of ourselves as intentional agents, that category distinction would not occur to us as significant, nor maybe even at all. So how can we be prompted towards understanding ourselves by encountering the spontaneous action of a 'natural' world that can only be thought of as spontaneously over against us on the basis of our self-understanding? How indeed do we recognise natural spontaneity without the *idea* of spontaneity? – and if we already have the idea, why should we need any encounter with external 'wildness' to put us onto it? How can we find external natural wildness helpful in grasping what lies behind our conceptual frameworks, when we must rely on those frameworks to conceptualise the affordances of the externally natural?

Actually, so far from these questions' threatening my argument, that argument indicates the only way in which we can make sense of the continuing back-and-forth jockeying, so well instanced by Soper, between 'realist' and 'constructivist' understandings.

Human beings have always looked out for guidance and inspiration to life from beyond the human, even though every way they have found to express this has necessarily been humanly – thus, culturally – framed. We can only resolve this by recognising that what we encounter as external wildness is *our own dark-selfhood externalised*. This recognition is the way to take seriously Schopenhauer's very deep maxim that 'we must learn to understand nature from ourselves, not ourselves from nature'.[14] We could not begin to see the non-human natural as 'acting from the whole of itself', unless we had something with which to contrast this – and that must be, acting from *less* than the whole, which we can only encounter as a possibility from the way in which we are drawn to alienate and tame *ourselves* in reflexive consciousness. We experience in such consciousness the inescapable occlusion of something that we intuitively need for making human sense. (The deep-structural and necessary – hence, universal – problem with consciousness turned reflexively on itself, as we have seen, is that its own living activity must always be beyond its grasp, and so it must inherently de-nature itself.) But that human self-experience is the only way that less-than-wholeness can *get into the world* – the world which is otherwise always and everywhere wholly what it is. (There could not be any *lack* in a world without reflexive consciousness, since things are only missing that are explicitly *missed*.[15]) Our self-experience of lack is therefore at the same time the only way that any contrasting conception of wholeness can arise, although by the very conditions of its arising it is something that we can only recognise in ourselves as reflected back from the world beyond us. When we construct external nature's indefeasible reality as 'wild', we are reconstructing our own wholeness out of the only materials in terms of which the structure of our self-awareness enables us to acknowledge it.

Then the existence 'outside discourse' of naturally wild wholeness is also the projection of an intuited precondition of our own sense-making. Its being *beyond* discourse reflects the *priority* of our dark whole selves to any discourse (any conceptual sense-making) that we can engage in. Any 'nature' we conceptualise must fall within the inverted commas, but its sensed underlying wildness and spontaneity reflect the intuited inner priority of our own wholeness to the creative–constructive business that we *do* with the inverted commas.

Wildness, to repeat, is the projected realisation of the dark self, our ungraspable wholeness read back from what we find answering to it in the outer world. We can sometimes, in art, catch this process in the act of happening: as when Yeats contemplates the swans in his poem 'The Wild Swans at Coole':

> Unwearied still, lover by lover,
> They paddle in the cold
> Companionable streams, or climb the air;
> Their hearts have not grown old;
> Passion or conquest, wander where they will,
> Attend upon them still.

This is so powerful because it is self-rediscovery, transcending self-pity – it is only in the spirit of 'my life is for me' that he has aged and wearied, but his whole self realises its wildness here undimmed. And significantly, in that rediscovery of wildness he can face death:

> Among what rushes will they build,
> By what lake's edge or pool
> Delight men's eyes when I awake someday
> To find they have flown away?[16]

This dynamic of dialectical realisation also explains the particular preferred forms that our symbolised wildness has always taken. Why mammals and birds but not insects, which swarm in their unguessable millions? We need, in projection, to be able to identify ourselves with what expresses our wholeness – and we are just naturally adapted to find certain things in experience attracting us in this way, as appealing to our deep need to make sense of ourselves (evolutionary pressure, for humans, cannot just be on our *material* being). That we can't do this with insects is a matter of the naturally given range of our sympathetic identification, which probably goes back to our inclination towards things with faces, or at any rate with forms and movements that more or less resemble our own. (Birds are at least bi-pedal.) Again, why rivers and lakes as well as the flora and fauna that are alive like us? These non-living things speak to us both as habitats of life, and also as manifestations of the power from beyond life that flows through it. (Water, for instance, as a constant image of permanent, unstoppable, always-changing force, reflects the fluid total responsiveness of the dark self to the emergent.)

But the kind of setting in which this life-beyond-human-ness is most congenially recognised will evidently vary with cultural and historical framings. (Hence Oliver Goldsmith's liking for both formal gardens and the 'music' of groves and brooks as contrasted with the barren emptinesses of Scotland,[17] but then not much later Wordsworth's 'visions of the hills / And souls of lonely places'.[18]) Nor can any particular form of it, for that reason, offer guidance or inspiration innocent of class, power relations and other kinds of context. We have to take what we deeply need in the forms in which our circumstances allow it to be available to us – as with the various satisfiers of any of our material needs, indeed. That wildness is an existential necessity – that we need active engagement with life from beyond the human, if our lives are to *be* fully human lives – is a general truth of reflexively conscious human experience as such. Although that necessity will always find expression in terms of particular kinds of natural environment, and out of particular cultural contexts, the general truth remains.

All this explains why I can make the kind of claim being put forward here, for the indispensable role of natural wildness in human life, without being committed to any of the positions on this topic that have (often justly) been found tendentious or otherwise problematic by proponents of 'inverted-commas-nature'. I do not have to suppose that nature is in itself and apart from human life ('intrinsically') valuable, nor that it is such value that somehow calls on us to harmonise ourselves with it, still less that there has ever been such a period of ideal harmony in any actual past.

I am excused, luckily, from wondering whether we should ever, under any circumstances, intervene in our biology. I don't need the implausible claims that the natural stands outside human power-relations and structures of interpretation (broadly, the 'ideological'), or provides us with any external *moral* criterion for action. I do not even need to argue that urban life is necessarily worse for us than rural life (as usual, it all depends), nor that rural life is not (in any period) a tissue of exploitation and class action, just like human life anywhere else. That wild nature is always 'nature' does not undermine, but actually explains, the kind of necessity that it carries for us.

... and hope?

What I *am* committed to claiming, however, and do actually claim, is that we cannot self-realisingly project the dark self in this humanly essential way into a *managed* world, one that now reflects back to us from its every facet the configurations of ego and conscious will. (Had Yeats had to write about 'The Swan Conservation Project at Coole' it would have had a very different ring – this is a theme that I try to take further in the next chapter.) It is the increasingly unavoidable stain of human will across the world – even with the best of progressivist intentions – that we really find appalling in melting glaciers, shrinking rivers, creeping desert and diminishing diversity. And we find this appalling *now*, not just 'for the future', though we recoil from its future consequences too, at both the prudential and ethical levels. But the tragedy here altogether precedes ethics. In such a human-configured world, laid under increasingly planet-wide contribution to improve the material human lot, we also increasingly lose our essential selves. Our human being wilts, and is parched and straitened, even before the outer desertification which we have triggered kicks in. This is no 'mere' metaphor but a profound existential truth.

The wild is the necessary *outside*, from which nature speaks back to us the living integration behind the ego that justifies and grounds it. Wildness is the objectified priority of that consciously ungraspable wholeness which we need for making human sense and for being undefeated by death. To see the human condition in that way makes for hope, because it is a necessary prerequisite of expectations that will not betray us – as any still-progressivist optimism is bound to do. But this isn't easy hopefulness, which would be no real advance on such optimism. It is the hope that can be learned (or rediscovered) only in acknowledging environmental tragedy: the hope that we can go deeper than our destructiveness – that *life in us* goes deeper.

We shall not easily retrieve a relationship of natural responsibility within a living world that we have now so extensively wrenched towards human purposes, and in the process have so radically jeopardised. But we may, once we know what the odds are, at least try to cope with the consequences of that jeopardy in that spirit.

Notes

1 Blackmore (2005), p. 3.
2 This and the immediately following passage come from Schopenhauer (1819/1958), Volume I, pp. 281–2.
3 See Lawrence (1962), Volume I, p. 360. The extended context of this phrase is very relevant to my whole argument.
4 From 'Auguries of Innocence', in Blake (1946), p. 118.
5 Montaigne (1580/1965) – see the essay 'That to philosophise is to learn how to die'.
6 See Foster (2008), Chapter 2, pp. 41ff.
7 Although this question is widely attributed to Forster, I cannot actually find a source for it. (If he didn't say it, he ought to have done.)
8 La Rochefoucauld (1967), no. 405 (p. 95) (my translation).
9 Keats (1954), p. 53 (Letter to George and Thomas Keats, 21 December 1817).
10 From 'Love' in Larkin (1988), p. 180.
11 Nussbaum (1990), p. 47.
12 The three extracts quoted are from Soper (1995), pp. 249–50.
13 See his essay 'Nature' in Mill (1874).
14 Schopenhauer (1819/1958), Volume II, p. 196.
15 The thought here is classically explored in Sartre (1943/1958).
16 Yeats (1919/1950), p. 147.
17 See Macaulay (1848/1906), Vol. ii, p. 443, quoting a letter written by Goldsmith in 1753.
18 *The Prelude* (multiple editions), Book I, ll. 465–6.

PART III
Retrieval

7

RETRIEVING WILDNESS

The theme of this last part of the book is retrieval: saving what can be saved of civilisation in face of the unprecedented human disaster that we have entailed upon ourselves and our successors. The considerations of Part II should have helped towards making clear what is really involved in this task, and why *retrieval*, with its suggestions both of *rescue* and of *repossession*, is the right word for it.

The implications of *rescue* alone, that is, will not suffice. Saving what we can of civilisation is certainly not a matter of preserving intact and functional as much of our present 'advanced' Western-style way of life as the climate consequences of having constructed that way of life will allow us to do. That would be to have learned nothing more from unfolding environmental tragedy than the technologies of obstinate stupidity. Nor, at the other extreme, is it a matter of working for the survival of enough communities of *homo sapiens*, in however diminished an economic and cultural state, to carry the species on as a biological entity through to the far side of climate catastrophe. Better extinction, surely, than that sort of desolate aftermath.

But nor, if we talk of retrieving culture, does this represent any kind of preoccupation with rescuing the Rembrandts from the rising flood-waters, and so on. Past achievements of art, order and civility obviously matter, as far as these can be retained, but what is absolutely crucial now is the re-establishment of a rooted *humanitas*: the sort of understanding and inhabiting of human being that goes with a recovered sanity of organic-conscious balance after the long pathology of progressivism. This balance is vital to any ideas of retrieval that could matter. It matters, indeed, more than the Rembrandts, because it represents recovery of the life-conditions for their creation, and so, if they must be lost, for fresh creation in some future where great art can once more be imagined.

Retrieval, in other words, means learning from environmental tragedy to recognise the essential human wholeness that contemporary progressive civilisation denies and thwarts. It also means repossessing that wholeness, as far as we can, in forms of practical living that can at least hope to endure through the consequences of that tragedy as they unpredictably emerge.

If that characterisation of the task sounds a rather grim note, it is meant to. The kind of hope being canvassed in this concluding Part is deliberately a very long way from what any contemporary politician or policy analyst is offering – or indeed is able to offer, given the discourse framework within which these people must take themselves to operate. The grounds on which we can call it *hope* are such that, if we seek to occupy them, the structural assumptions of that discourse are fairly comprehensively subverted. Such hope challenges us at a depth from which politicians' talk of human purposes, not just in terms of 'winning the global race', but also in terms of apparently more intellectually respectable goals such as economic growth, social progress and even sustainable prosperity, can be seen as so much futile chatter. If this book is anything like right, however, they are for all that the only grounds of real hope, and retrieval will depend on this recognition. I shall therefore pursue my hunch that the book is indeed right, or could well be right, onwards through reflections that may seem unpolitical towards conclusions that will almost certainly be found unpalatable, in the spirit of that hope.[1]

Retrieval, if it can be achieved at all, will be a long, complex and exacting task, and in the next chapter I touch on some of its key practical aspects from the particular angle represented by this book. That angle involves emphasising that as part of being a practical task, it is a *philosophical* task. Unless we rise to the challenge of serious rethinking, our responses to the exigencies which increasingly break upon us will be defeated by a desperate continuing progressivism reinforced by more and more sterile and destructive forms of denial. As I have tried to show, these attitudes, or casts of mind and spirit, lie very deep. Freeing ourselves from them means retrieving for our lives a habit of intuitive natural responsibility – and that in turn means, as I hope the last chapter has made clear, restoring to those lives some real connection with the wildness that is at the root of them. That is not something to be done only or even mainly in the policy domain, but first – and crucially – in that of renewed self-understanding.

Done in that domain, we do not know how it will strike or tell, with whom or to what effect. Effectiveness here, indeed, is not a matter of majorities and falls well outside the realm of the predictable. But unless something along these lines is done, then as the gross failures of the last thirty years of environmental policymaking (or non-making) come home to roost, we will have no viable framework for the genuine hope that might carry us through. It falls now above all to a realistic environmental movement, one that has at last stopped pretending, to make sure that we are better placed than that for encountering whatever is coming.

The philosophical dimension of the retrieval task cannot be separated from how we tackle the practical exigencies. Across the whole range of political, communal, economic and personal activity, we will need to free ourselves from long-entrenched ideas and patterns of thinking. That is why in the next chapter I approach some of the key areas of concern by offering, not policies, plans and prescriptions, but a basic conceptual toolkit for asking the right questions. As a prelude to this kind of critical engagement, however, we need from the present chapter a preliminary sketch of the larger picture. What, in the context of oncoming climate instability and major ecological damage, could *retrieving natural responsibility* actually mean? What could putting our lives back into living touch with their fundamental wildness really involve? What are we thinking about here, that could possibly be adequate to the adversity that is impending?

Rewilding?

One temptation that has grown stronger of late is to suppose that we are thinking about 'rewilding': the project of allowing ecosystems space to recuperate themselves naturally, with human pressures deliberately taken off them, and associatedly of deliberately reintroducing various species – some of them quite startling choices – to habitats from which human activity has driven them over recent decades, centuries or even millennia. This project arises as a very understandable response to our current alienation from the non-human natural world. I want to begin, however, by differentiating what I have in mind quite sharply from it, in order to bring out by contrast the much more startling character of what genuinely retrieving wildness must entail.

The environmental activist and journalist George Monbiot is (as I have already remarked) one of the few who have so far been willing to recognise the end of pretending. In his *Guardian* column written shortly after the Rio+20 summit in 2012, he accepts that the failure of even that comparatively unambitious jamboree marks, more or less, the end of the multilateral effort to protect the bio-sphere ... that we have missed the chance of preventing two degrees of global warming now seems obvious. That most of the other planetary boundaries will be crossed, equally so.

To the inevitable question: in that case, why go on campaigning? he offers three answers. In the first place, we can draw out the losses over as long a period as possible for the sakes of our children and grandchildren – so that they can experience at least something of 'the forests, the brooks, the wet-lands, the coral reefs, the sea ice, the glaciers, the birdsong and the night chorus ...', before all this is laid waste. Second, we should go on attempt-ing 'to preserve what we can in the hope that conditions might change' – which is at least not impossible, although he cannot honestly foresee it happening. But, third, we can do more than simply preserve. 'Rewilding – the mass restoration of ecosystems – offers the best hope we have of creat-ing refuges for the natural world.'[2]

As good as his word, Monbiot himself has plunged wholeheartedly into this third option, and by a year later had produced an interim report on his ideas and investigations in his latest book *Feral: Searching for enchantment on the frontiers of rewilding*. Here he redefines the term 'rewilding' (which apparently entered the dictionary only in 2011) so as to extend its signifi-cance well beyond its original meaning of releasing captive animals back into the wild. For Monbiot, the idea is

> about resisting the urge to control nature and allowing it to find its own way. It involves reintroducing absent plants and animals (and in a few cases culling exotic species which cannot be contained by native wildlife), pulling down the fences, blocking the drainage ditches, but otherwise stepping back. At sea, it means excluding commercial fish-ing and other forms of exploitation. The ecosystems that result are best described not as wilderness, but as self-willed: governed not by human management but by their own processes. Rewilding ... lets nature decide.[3]

In pursuit of this vision he discusses a variety of rewilding projects, viv-idly records his own engagement with wildness on sea and land, explodes some conservation myths (such as the value of artificially propping up hill farming) and speculates on the possible reintroduction of creatures such as wild boar, lynx and even wolves to the open British countryside. It is an exciting, in many ways indeed an enchanting book, and I would press it on anyone interested in the issues of humanity's relation to wildness.

But what about global warming? Monbiot even canvasses, on the wilder shores of rewilding, the reintroduction of free-range elephants to Europe, but meanwhile climate change is already something of an elephant in the room for his position in *Feral*. It receives its first explicit mention only on p. 125, and that is just in a passing reference to causes of past extinctions. All the other references in the book are of the same order – comments on present or past climatic shifts as factors among others affecting species survival or rewilding prospects. Nowhere is there any confrontation with what might be called the 'deck chairs on the *Titanic*' question – what could be the point of rewilding a planet *en route* to atmospheric temperatures of 6°C above pre-industrial levels of CO_2?

If in the light of that question we return to his *Guardian* article and those three justifications for continuing activism, it is surely clear that we can diagnose much of their inspiration as a form of interpretive denial. Yes, we can turn our immediate attention to preserving what we can for our grandchildren and to creating areas of 'self-willed' ecosystemic renewal – but isn't this really just to change the subject, to turn our eyes away from the implications of medium-term futility attaching to any such enterprises in our imminently overheating world?

Maybe this is not the whole story. There is a subtext in *Feral* that connects its downplaying of anthropogenic climate change with Monbiot's second justification, the hope that conditions might yet unforeseeably change. 'We need to show where hope lies', says an impressively single-minded rewilding activist whom he quotes at one point; 'Ecological restoration is a work of hope.'[4] Instead of telling people what they can't and mustn't any longer do, because of forces beyond their control (something which environmentalists have always been sternly good at), this work of hope shows them positively where new opportunities and freedoms may lie. If people then start to hear ecological restoration as speaking to what they find deeply missing from their lives, they might become more accepting of the carbon-related constraints that are a necessary condition for its having any lasting point. And engaging with such restoration might bring back into play a sense of habitable wildness which could maybe generate (as sustainable development discourse so palpably doesn't) a wider revulsion against climate-wrecking forms of living. At this level, the rewilding impulse could be seen as expressing neither interpretive denial nor facile optimism, but the kind of hope that can admit and still be undefeated by tragedy.

Perhaps. But if this is indeed the level at which it is trying to work, the tensions that are visible within the rewilding approach as Monbiot sets it out become correspondingly serious. These tensions will have been glimpsed even in his definition of the process as quoted above: we have humans on the one hand 'resisting the urge to control nature', but on the other hand intervening to cull a few 'exotic species' here and there. (Surprisingly, these include sheep, at least in the UK – if you thought sheep were too sheep-ish for anyone to dislike, Monbiot is the man to disillusion you.) Tension becomes acknowledged contradiction when he celebrates rewilding as addressing our own unmet need for a wilder life – in an intriguing chapter he ascribes illusory 'big cat' sightings around Britain to 'an unexpressed wish for lives wilder and fiercer than those we now lead. Our desires stare back at us, yellow-eyed and snarling, from the thickets of the mind.'[5] Yet at the same time he insists:

> It should happen only with the consent and enthusiasm of those who work on the land. It must never be used as an instrument of expropria-tion or dispossession ... Rewilding, paradoxically, should take place for the benefit of people, to enhance the world in which we live, and not for the sake of an abstraction we call Nature.[6]

But to stipulate consent and endorsement in this way is to control our very stepping back from nature, and thus to limit its 'uncontrolled' reassertion of itself ('Rewilding ... lets nature decide') to the areas where we feel com-fortable with it and expect to benefit. He is obviously aware of this tension, as witness his 'paradoxically', but he doesn't face up to it – and yet, surely, such conditionality brings the rewilding impetus back to the level of little more than a quality-of-life aspiration.

The difficulty here is most evident in what Monbiot writes about the prospect of reintroducing wolves to Scotland. After glancing at various ways in which their restored presence could (so he argues) be of service to conservation and the local economy, he goes on with admirable frankness:

> [I]t would be deceptive to claim that I would like to see wolves rein-troduced because they kill foxes or reduce disease or assist the owners of grouse moors and deer estates. I want to see wolves reintroduced because wolves are fascinating ... because they feel to me like the shadow that flits between systole and diastole, because they are the nec-essary monsters of the mind, inhabitants of the more passionate world against which we have locked our doors ... But it should only happen if there is broad public enthusiasm for the project.[7]

This is poignant, and moving in its very conflictedness. For all that, it should be evident that necessary monsters for which there is broad pub-lic enthusiasm aren't going to be *real* monsters, even of the mind. The wolf at the door for which, as a prerequisite of such enthusiasm, convincing management arrangements would need to be in place, remains safely locked out. Thus at the point of becoming drastic enough to worry anyone, rewilding is revealed as a kind of elaborate game. It shares, in fact, the fundamental unseriousness of pornography or the horror movie. We play at ceding control, stepping back just far enough to be able (if we're careful) not to see ourselves still running the show. Rewilding turns out to be a way of *not* getting beyond pretending, after all.

This is not to deny that any 'managed' re-wolving of the Scottish Highlands, for instance, would still involve danger – simply because we are not very *good* at managing complex living situations (that is one main reason, after all, why techno-fantasy à la Lynas *is* fantasy). There would always be genuine wildness lurking within the situation, just to the extent that it contained this permanent likelihood of our managerially messing up. But we should nevertheless be trying hard to minimise that danger, and that cast of mind can't count as welcoming the wild; we only encounter wildness in the right spirit, to the extent that we recognise danger as *savingly inherent* in what we are doing. (In the brilliant line of Hölderlin: 'Wo aber Gefahr ist, wächst / Das Rettende auch' – where there is danger, thence arises also salvation.[8])

We might also consider here this, from a more recent *Guardian* column by Monbiot:

> Why shouldn't every child spend a week in the countryside every term? Why shouldn't everyone be allowed to develop ... rock climbing, gorge scrambling, caving, night walking, ropework and natural history? Getting wet and tired and filthy and cold, immersing yourself metaphorically and literally in the natural world: surely by these means you discover more about yourself and the world around you than you do during three months in a classroom.[9]

I am not inclined to underrate the educational importance of such contact, or the ways in which it can develop personality and enrich imagination: and the degree of psychological and imaginative benefit, even from controlled exposure, will be the greater the more deprived in this respect those exposed have previously been. As Monbiot points out, it is a marker of how far alienation has gone that 50 per cent of children in the UK have never visited the countryside (at least, according to the adventure learning charity Widehorizons[10]); or that some of the children whom he met before writing his article had never seen the sea, and *none* of them (he claims) knew that a nettle stings you. But it is only to the extent that we expose children – or anyone – to real wildness in such contact, which means real risk rather than managed, risk-assessed risk, that the *existential* benefits are going to be forthcoming. And could we imagine ever doing that with children, as part of their education, except in a thoroughly controlled way?

'Feral is what children should be', says Monbiot, teasingly but cogently; 'it means released from captivity or domestication'. Its root meaning, however, is *un*domesticated rather than escaped from domestication – still less *released*, which won't be unconditional. I am far from saying that people – kids or adults – get nothing from trips to the countryside under controlled conditions. There is always the background of dangerous possibility: people sometimes die falling off Striding Edge, a route along which thousands toddle each year, and the awakened awareness of even possible 'outsideness' is worth something. But it is not enough when *retrieval* of the kind about which I am talking in this chapter is at stake. We must be genuinely exposed, beyond any game and without safeguards or reservations, to be confronting wildness worth the name – and, as I have argued, we need such wildness for our lives to be lives. Where will we find that exposure in a world that humans seem to have irretrievably overrun? Out of what darkness might now come the wolves of unmanaged otherness that we do so vitally need?

Climate beyond control: the conditions of retrieval

It is time to redeem the pledge at which I hinted in the Prologue, and to consider how we can think positively about the enormous natural forces that humanity, in triggering global climate instability, has released – but has not generated.

If we are honest about the conditions that will be imposed on any retrieval of civilisation in face of these forces, we might sum them up under these three closely connected heads. First, *we don't know what's coming when.* Second, *we aren't going to be in control.* And third, *progress is over.* These truths summarise and bring together much that has been discussed in earlier pages, so I will here just recapitulate and underline the most important aspects.

We don't know what consequences and possible non-consequences of humanity's long 'progressive' carbon-binge are coming when, simply because for all our scientific sophistication there is too much ineliminable contingency in the system. We are all tacitly aware of this just as lay people, whatever airs scientists and techno-managers may put on, because it is so clear in our ordinary life-experience that very rarely do things turn out as predicted in even the simplest matters. But there is also some more technical terminology available in which to capture this recognition.

Thus we can refer to the social contingency of science – the fact that precise, determinate numbers both inherit and conceal the negotiated, essentially open-ended nature of the processes through which they must be established. Accurately assessing the numerical value of any parameter requires not just appropriate instruments but a whole system of institutionalised collaboration, organised and disciplined by agreed *criteria* for such things as sample choice, sample manipulation, instrument calibration, methods of data recording, training of personnel and controls for inter-laboratory bias. The point about such criteria is not that they rob the resulting measurements of 'objectivity', which actually can be achieved by no other means, but that they depend on collaborative judgement, which in turn depends on practices of attention, selection and choice. The numerical values that emerge may be precise, but they will have been *given* their precision through tacit framing by these collaborative and inherently negotiated processes.

But then the underlying indeterminacy here is hugely multiplied by the need to feed such numerical values into scientific modelling of future trends and developments in complex living systems. Real-world natural systems are multi-variant and emergent – they are both interactive and open-ended, such that future system change cannot be determined from any given system state. Predictive scientific models, on the other hand, are necessarily selective, representative and finite. The numbers that they emit will therefore inherit not just the ordinary contingency of any scientific process, but also that flowing specifically from the need to make judgements in model-construction which attempt to capture the radically non-linear relationships found in living nature, where when one variable shifts another may change exponentially – and to project these relationships ahead over seriously extended time spans.

Of course this doesn't mean that we can't predict environmental and climate futures *at all*. At a sufficiently low resolution, there are things in prospect about which we can be fairly confident. We know, for example, that world population is likely to go on increasing, and (bar apocalyptic accident) to pass the nine billion mark before stabilising (if it does stabilise), though we don't know just when or by how much it will pass that mark. Similarly, global mobility and the pressures of techno-agriculture are likely to produce more pandemics of the swine flu variety, though each particular outbreak is likely to take us as much by surprise as that one did. And there is now a huge expert consensus behind the conviction that greenhouse gases generated by human activity are indeed warming up the world. We should not have known this last, vitally important truth without the science. However, there are so many interlocking ifs along the way to computing its consequences – *if* the Arctic ice disappears, *if* the Gulf Stream alters direction, *if* Amazonian tree-cover is reduced by vast forest fires, *if* the capacity of the oceans to pump down carbon dioxide collapses ... – that any offer at precise prediction will have little more epistemic value than confessed uncertainty.

Hence, of course, we are lacking a fundamental requirement for being in anything like real control of whatever may be coming. Again, that doesn't mean that we can only respond blindly or after the fact. As we shall consider further in Chapter 8, there is an extremely important role for informed precautionary anticipation, and for developing the sort of resilience which it can foster. But the precisely targeted geoengineering manipulations and smart bio-interventions beloved of the techno-fantasists are simply out of reach in this area, because – again, if we are honest – we must admit to something very close to ignorance about anything more than their immediate real-world effects, and sometimes even about these. (The history of genetically modified crops and the controversies over their possible unintended consequences provide abundant illustration here.)

We are liable to mislead ourselves in these matters by relying on a very selective image of our technical capabilities – the selection, in fact, might well be seen as yet another form of interpretive denial. We can surely make a success of geo- and bio-engineering, we tend to feel, because we (or, a representative few of us) can point a rocket at Mars, some thirty-four million miles away at its nearest, and after about eight months the rocket often gets there. This is the kind of thing (characteristically involving missiles) that technological humanity has become quite passable at. But controlling the oncoming consequences of our impacts on natural systems, with their dynamic open-ended complexity and multiple feedback loops, is like a vastly more difficult version of trying to manage the world economy – something at which considerable experience ought by now to have convinced us that we are rubbish. Here, as there, precaution and informed resilience really represent the best response to the unpredictably emergent that human beings can hope for.

And since the continuance of progress – conceived as material standards of living continuing to rise overall despite the gathering stringencies of a warming world – would clearly have to depend on much more specific prediction, management and control of increasingly severe climatic and ecological consequences than that, we have no option but to accept that progress is over. Although wishful optimists are now (as we have seen) locked into anticipating such control, it is clearly a fantasy. As with peak oil, so with peak progress: it is dictated by factors that are now beyond our ability to influence significantly, and it is nearly upon us.

In view of all that has been said about it in earlier chapters, the ending of material progress as an *ersatz* goal of life is not, if we are coldly honest, something to be mourned. But we have also to reckon with progress*ivism*, the pathological attachment to perpetual betterment which I diagnosed in Chapter 3. Material progress stalled or reversed while still craved by populations and electorates across the world as the principal human good, will create an explosively dangerous geopolitical situation, and this too must be factored into our recognition of the oncoming instabilities and uncertainties that we shall increasingly be unable to control.

The human condition

Such a prospect must represent intolerable human defeat, unless we can learn to see it in a different light. But now, after the work of the last chapter in clearing some conceptual space, we may begin to recognise these fundamental constraints on retrieval as expressing three directly corresponding principles of wildness in all human action: we can't *ever* be sure what will happen, we are *never* in control and there are *never* any guarantees. If we can understand how these principles express the inescapable nature of agency in *any* kind of situation, what might have felt like helplessness in face of the climatic and ecological forces that we have triggered might turn into something very different: a renewed acknowledgement, after long scientistic and managerialist delusion, of the reality of our human condition.

This acknowledgement, if we could reach to it, would offer a vital re-encounter with wildness. We can only explain the generality of those conditions by appeal to the way in which action must proceed from natural responsibility. That involves an intuitive awareness of what I have called the dark self, the always-tacit locus of our whole organic-conscious being, if we are to make habitable sense of our activity as agents. Understanding these conditions as principles of human action shows us how we need to listen for our wildness if we are to understand ourselves at all.

The claim that we cannot *ever* be sure what will happen will be recognised by those with some philosophical background to hark back to the great Scottish Enlightenment thinker David Hume, who had a very well-known argument to this effect.[11] All we have to go on in prediction, Hume pointed out, is experience – our own, and that of others – in either the present or the past. We naturally and instinctively take it that rules and scientific laws generalising from that experience will go on applying, and so we predict more or less confidently on that basis. But if we chance to ask *how we can be sure* that rules based on past experience will go on applying to future experience, the only answer we can produce is that they always have so far: and evidently, that is worth precisely nothing as an answer to just *that* question. So, concludes Hume, in principle we *can't* ever be sure.

This has its inescapably paradoxical air because clearly we do assume that the future will be governed by broadly the same laws as the past, and we so often trust life and limb to that assumption (every time we get into a car or a train, for instance) that to call it doubtful seems entirely perverse. We just know that we can trust the assumption. But it is the whole organic-conscious self, always dark to our conscious thought, that tacitly trusts it and presses forward into the future, borne on the current of instinctive life. Any living thing that did not live in such instinctual trust, if we could imagine one, would be quite paralysed for life-action and very rapidly dese-lected. Yet in the pure light of reason (that is, in the space of sufficient reasons subtended by the subjective ego-self set over against its objects), this life-essential assumption can only stand unproven, apparently hazarding us afresh at every tick of the clock.

It is in fact another, and a classic, example of a paradox in which we live daily but find habitable only out of the intuited reality of the dark self behind the 'I'. Finding it habitable, however, doesn't mean that the paradox is simply dismissed. Rather, it remains always potently active, alerting us more or less subliminally to the leap into the unproven, the bet on the emergent, which is always being made in even the most routine prediction, and putting a premium on sensitivity to those more demanding cases where the decisively new may well require us to revise the laws and generalisations that we have based on our experience hitherto. The whole self operating at full alertness moves always at least slightly warily into each new moment – and the worst cases of mental taming come whenever we lose that sensitivity, that essential wildness of epistemic habit, and begin to constrain the reality of our actual experience to fit the interpretive paradigm in which we have invested our comfort and security.

As you would expect, our *never* being in control is a closely related matter. Control requires not just prediction but also planning: to be in control we need to know not just what is coming but how we are going to react to it. But planning what we are going to change in the future can only go on within a framework of assumptions about what will hold firm. This, however, generates another paradox of the same order. When we plan within a framework of such assumptions and then, in order to stick to our plan, hold that framework constant in the face of emerging events or information that suggest changing it, we are not in control – we can't be, since our assumptions no longer reflect reality. But if we change the framework to accommodate the emergent, we are not planning so much as responding to events – so again, we can't be in control. So, we are never in control, because *any* plan must make framing assumptions, which are always therefore going to be exposed to the emergent at one step ahead of our planning. And again, we can only inhabit what here confronts the ego-self as a paradox of control by deploying, intuitively from the whole self, a kind of tact of living engagement where the plan rolls with the punches that are directed according to the plan. We go on coping by accepting that we can never really steer the wave of circumstance, and must therefore ride the breaking crest of *now* as skilfully as we can manage.

Control, moreover, is not simply about prediction – we can, after all, seek to predict without seeking to influence in any way what we predict. Control is about using prediction to ensure that what happens conforms to our will: it is about knowing both what *will* happen as a result of our doing A or B, and what we *want* to happen. But just as predictability might fail in relation to the upshot (I can never be sure, for instance, how a new group of students will react to an exercise that has worked well before), so might self-knowledge in relation to my will. (Did I, after all, really want to get the same response again? – or do I now begin to understand the purport of the exercise differently?) Any complex enterprise is a matter of finding out whether one really wants what one embarked on it to attain – 'How do I know what I want until I see how I feel about what I get?' (Of course, one can always discover that, yes, that *was* what one wanted – one can be reconfirmed in one's intentions even as one rolls with the punches of implementation.)

The self whose relation to events is to be one of control has to know the world of action predictively, and its own desires intimately from within, unqualifiedly as it were, and this sits well with the model of subjective self over against objective world. If 'my life is for me', my life is for getting what I want out of a world organised by the principle of sufficient reason. To aspire to be in control, one has to construct oneself as an ego. But as an ego I can never be in control of my real wants, any more than I can of my working assumptions, without getting caught in the kind of paradox just explored: if what I want doesn't flex with what I find I can get, I am always liable to be willing the world into a false mould, and this is not to be in control of anything – but if what I want does so flex, it immediately looks as if I am *being* controlled. We have to work with a tacit model of both assumptions and wants as emerging from the dark self: what, at bottom, as a whole responding self, do I ongoingly *find* myself committed to thinking and wanting? By the same token, our basic desires and organising assumptions have to 'grow wild' – if they are cultivated (that is, explicitly and deliberately chosen) they are not basic, and an obvious regress starts up. At the roots, desire has to grow in the dark of the unattended self.

That there are *never* any guarantees is, once again, an intimately related point. This hard truth of the human condition re-emerges decisively from the ending of material progress, attachment to which has tried for so long to build into the ordinary forms of life an insurance against the future: whatever happened, things overall would only change for the better. As we have seen, the pathological strength of this attachment to progress arises ultimately from the attempt of the ego-self to protect itself against recognition of its own finitude, which it must acknowledge but cannot grasp. Maintaining an uninterrupted trend of general betterment is then the only context in which the project of having projects can appear plausible. But the aspiration to guaranteed material improvement is not just something which events have now tragically defeated, as a consequence of our progress-driven technological pressure on the ecological limiting-conditions of life on Earth. It also, just *as* an attempt to insure against the future, travesties our necessary relation to the always-novel emergent within which our conscious agency must be set. D.H. Lawrence puts a blunt expression of this vital truth into the mouth of one of his most uncomfortable personae, the more-than-gamekeeper Mellors in *Lady Chatterley's Lover.* 'You *can't* insure against the future, except by really believing in the best bit of you, and in the power beyond it.'[12] It is not just that chance won't ultimately be tied down, so that all we can ever finally trust to is the best we can do from moment to moment. As a corollary of that, and in the light of our radical lack of control, life is always radically open. This is a truth of spiritual ecology directly related to Aldo Leopold's famous paradox of conservation: 'too much safety seems to yield only danger in the long run … In wildness is the salvation of the world.'[13] To the extent that we focus on trying to guarantee, *really* guarantee, some future outcome, we disorder current living, of which the essence is responsive adaptation of our understanding and criteria in the face of unanticipated change – and so we actually rob ourselves of both the present and the future. The pursuit of any guarantee against deterioration in our life-circumstances is thus always inherently self-cancelling.

Once again, however, we can only properly understand what is being pointed out here by referring it to our intuition of the dark self. Really believing in the best bit of you means living from it, and that is something that makes no sense from the perspective of the conscious ego. As an ego-self, whatever powers and capacities I may contingently have are 'for me', and so even the best of those endowments is always being instrumentalised to serve the fundamentally most valuable thing, the undifferentiated me-centre. Nothing here corresponds to any 'power beyond', only ever the ego at the heart of things over again. But 'the best bit of myself' implies a view of myself that is not thus first-personal, yet nor can it be third-personal

– because then the issue would arise of whether and why I accepted, subjectively, the criteria of 'best-ness' being used. Rather, it must come out
of the intuited fullness of living being. In actual inhabited sense, the best
bit of you is that aspect of your gifts and capacities, trusting to which
best expresses your life as the whole that is dark to self-consciousness and
ungraspable – the whole that integrates you most viably as an embodied
intelligence making its particular unique way through the world. If we
live thus creatively, we touch our essential wildness (the creature's living
untrammelled from the full of itself), and find the source of power from
beyond that really informs and nourishes the will.

In these thoughts about the human condition we are being brought up
against the radical difference between *psychological* and *existential* resilience.
This is something on which I touched briefly when discussing denial in
Chapter 1, and it has been in the background while we have been considering the structure of the self in Part II, so it will be as well at this point to
deal with it explicitly.

Psychological resilience is the capacity of the ego-self to respond to stress
or shock without denial or disabling anxiety. Denial here would be hiding
from ourselves parts of what we really know, and anxiety a sense of the
ego-self's associated contingency, its brittleness as a centre. Against these
threats, the inner strength of psychological resilience is a condition where
the centre *holds*, against stress or shock: potential disintegration into denial,
or indeed into madness (functional breakdown of the self), is resisted from
within. Resilience here is the kind of strength 'for me', for keeping myself
intact, which my inner life can be experienced as having – or can, to some
degree, be trained to acquire. But existential resilience is precisely *not* about
'inner life', though it is about experienced being. It comes not from deeper
in, but from beyond (and what is beyond the centre cannot be deeper in,
since only the deepest in can be central). It is in fact the capacity to cope
with the stress involved in our necessary exposure to *being an ego-self*: that
is, to the specifically human, reflexive-conscious condition. It is resilience
not against the attacks of circumstance on the gathered self, but against the
strains on wholeness of being which that self inevitably brings with it. It is
the strength to *be*, despite having to be human *as* an ego-self.

Recognising the relation between the inevitably centred self and its
being-beyond-itself – that which I have been calling the dark self – is then
in some sense like religious awareness, though it consists not in any formal
rituals, but in a sustaining sense of the beyond always actually or potentially
informing the immediate. It is the intuitive knowledge that comes to Tom
Brangwen in *The Rainbow*, watching by the sheepfold under the flashing
stars: 'He knew he did not belong to himself'.[14]

Retrieval: general guidelines

If we can indeed allow ourselves to be reawakened by oncoming climate tragedy to these permanent fundamentals of the human condition, and the essential wildness to which they make appeal, we are already rebuilding that vital existential resilience. We may then find it easier to accept some of the guidelines for action deriving from those three constraints on retrieval indicated above. These were, to recall: that we don't know what is coming when (that is, more than very roughly the extent or timescale of climate effects); that we are not going to be in control, either of these effects or of much that will happen as we respond to them; and that, consequently, our settled expectation of material progress must henceforth be abandoned.

Before trying out some conceptual tools for thinking about action under such conditions, therefore, I will identify briefly the derived guidelines of this kind that seem most relevant.

In the first place, as already noted in the Prologue, we must get used to there being no blueprints. There is no producible plan or policy framework for retrieval analogous to those that used to get brandished in the earlier days of environmental engagement with mainstream politics – of which well-known titles such as *Blueprint for Survival* or *Blueprint for a Green Economy* were characteristic expressions. This is not just because such plans have all too often been excuses for established elites to capture issues of concern as 'problems' too big and complex for local or individual action, in order then to address them (or fudge them) only on the elite's own terms – as in the case, most dramatically, of 'sustainable development'. It reflects, more fundamentally, the condition just explored: that under the kind of uncertainty that we now face, any blueprint must necessarily falsify, leaving us not more prepared at all but actually more vulnerable.

Second, nevertheless, we need to encourage ourselves with the guiding thought that as long as we don't expect too much (a habit of which seeking blueprints is in this context a clear illustration), we can often get by. As has been well said, 'Life is full of trouble, much of which never happens'. Just as it is often possible in ordinary affairs to achieve some success if one is not hell-bent upon victory, so we may be able to find ways to survive even in this grimly tightening situation for quite a long time – and *maybe* for long enough to see the tide turn – if we are no longer locked into the non-negotiable demand for progress, or even for non-regress. Once we can see getting by on those terms as a matter not of grievously lowered expectations, but (in our real circumstances) as one of sanguine though not yet wholly unrealisable ambition, we may be able to go on in hope.

Third, however, such hope will have to consort with the recognition that everything is risky, and will get increasingly riskier. As a corollary of that, we simply have to be prepared to stand the damage. We will only survive by trying out things that will inevitably produce casualties, as well as possible ways of working onwards. Taking all policy options to be constrained by the requirement that, at least in peacetime, no one gets killed or seriously hurt, is another form of pretending which we will now need to move explicitly beyond.

A further guideline implicit in these three – though perhaps less obviously – is to cultivate the learning virtues in all we do. We have already glanced at these virtues in the previous chapter, as life-strengths on which we depend to make any headway as organic-conscious sense-makers. They include *criticality*, which responds to our perpetual temptation to shirk the emergent, to accommodate the new to the familiar, to domesticate potentially challenging experience under concepts with which we have grown comfortable. Criticality is the life-strength to resist this – to be always watchfully ready to challenge our taken-for-granted assumptions, if only to ensure for them that permanent instability without which they so easily solidify and jam up our being. Closely associated with it is, to coin a term, *explorativeness*, the robust persistence in canvassing and testing possible interpretive frameworks of assumption alternative to the currently taken-for-granted, even before one is moved to adopt any of them in its place – a multiplication of available cognitive options that always increases the epistemic value of one's already-learned understanding. This is analogous to the way in which building real options into a capital asset – the technical possibility of an engine's switching between different fuels, for instance, according to changing market price – increases present, rather than discounted future, asset value.[15] The epistemic equivalent of such present value might be a reduced temptation to defend and insist on one's current learned understanding in the face of emerging experience that it can't really accommodate.

We might also add sheer *epistemic wildness* – the readiness and resolve to mix it sometimes, just to see what happens. This involves not merely enduring uncertainty when interpretation is difficult or ambiguous, but actively courting it as a potentially creative matrix for deeper insight. In its own modest and tentative way, the present book has aspired to exhibit something of that readiness.

Together these virtues define a characteristic disposition towards the future – a kind of focussed attentiveness in which we deliberately let the unanticipated inform us out of experience. Doing that involves a self-possessed and resourceful waiting on what transpires. Only so can we conform to the last and perhaps most important guideline for retrieval: don't try to look too far ahead. As I remarked in the Prologue, this is likely to be the hardest injunction of all for us to follow, products as we are of at least half a century during which 'predict-and-control' has been our main political watchword. But the conditions under which we must now retrieve what we can, mean that any anxious looking to the longer term will tend only to betray us. The crucial thing we have to learn from the end of progress is to divorce hope not just from false utopianism but from any kind of ideal goal. The only hope that now stands any chance of carrying us through is a non-religious form of that faith that Tolstoy so compellingly identified as the strength within us of life itself.[16]

These broad guidelines for retrieving what we can from oncoming climate turbulence and ecological disarray, in a spirit of existential resilience informed by natural responsibility towards our intuited wholeness, express that kind of hope. Bearing them in mind, then, let us turn to some more focussed thinking about some of the contexts of action.

Notes

1 'Winning the global race' offers a uniquely fatuous version of the project of having projects, one that will not survive a moment's thought (what could count as victory in this race? what happens afterwards?) – yet it is a metaphor routinely invoked by the British Prime Minister at the time of writing (David Cameron) as if it actually meant something, and none of his opponents calls him to account. This small indicator of the helpless vacuity into which our progressivist politics has sunk should be borne in mind if what follows begins to seem just *too* grim, or too impracticable.
2 The column from which all these quotations come appeared in *The Guardian* for 25 June 2012, as 'After Rio, we know. Governments have given up on the planet'.
3 Monbiot (2013), p. 9.
4 Monbiot (2013), p. 152.
5 Monbiot (2013), p. 60.
6 Monbiot (2013), pp. 12–13.
7 Monbiot (2013), pp. 117–8.
8 Hölderlin (1951), p. 165: from his hymn 'Patmos'.
9 'The problem with children? They just aren't feral enough', in *The Guardian*, 8 October 2013.
10 See www.widehorizons.org.uk
11 See Hume's *Enquiry Concerning Human Understanding* (multiple editions), Section IV.
12 Lawrence (1928/1960), p. 313.
13 Leopold (1949), p. 133.
14 Lawrence (1915/1997), p. 38.
15 See Foster and Gough (2005), Chapters 8 and 9 in particular.
16 See Tolstoy (1921), p. 60 (from *A Confession*: 'Faith is the strength of life').

8
TOWARDS A TOOLKIT

In this chapter we finally come as close as a book of this kind can come to the *practicalities* of retrieval. But the book's kind is avowedly conceptual, so how close is even that going to be?

What follows is not a 'handbook for retrieval', nor even an extract from such a handbook, for two reasons. First, the practicalities of retrieval responses will be enormously multiple and various – different in different communities and localities, dependent on where they are starting from, what resources they dispose of and what capacities they can deploy. Even to begin tackling such detail would be a task requiring very many more thousand words than I have left for this book. It would also need to call on economic, sociological, political, organisational and agricultural expertise to which I cannot pretend.

The second reason for its not being a handbook, however, indicates why this incapacity may not be a drawback. As already emphasised, there are not going to be any blueprints here. There will be no plan Z, of the sort envisaged by Thomas Homer-Dixon: 'detailed scenarios of plausible climate shocks; close analyses of options for emergency response by governments, corporations and non-governmental groups; and clear specifics about what resources – financial, technological and organisational – we will need to cope with different types of crisis'.[1] We are moving into a world where that kind of control of events by governments, even with scenarios and options built in, will be a thing of the past. But nor is there really going to be any reliable equivalent at the level of community initiatives and local resource-building. This is a difficult point to make, because if there is nothing like this to be relied on at either higher or lower levels, what prospects for anything which we could recognise as *coping* with this crisis do we actually have? It is precisely here, however, that the importance of having cleared conceptual space for what I have called *existential resilience* becomes apparent.

I say more about the relation of this idea to those of economic and community resilience below, and in the process (I hope) bring this book helpfully to bear, in its own mode of helpfulness, on the practical challenges which on-the-ground retrieval will face. But that mode is deliberately different from any handbook. In this chapter I am trying to loosen the conceptual binds from which we will need to have freed ourselves in order to create the conditions that are necessary to retrieving what we can, in a context where blueprint-thinking (and to some extent, handbook-provision) may still mislead and betray us.

It is by way of insisting on the ultimate practicality of this work that I call what I offer here the beginnings of a conceptual *toolkit*. This metaphor is borrowed, with thanks, from Daniel Dennett's book *Intuition Pumps and Other Tools for Thinking*.[2] He there describes thinking tools as 'handy prosthetic imagination-extenders and focus-holders', classifying them into a variety of kinds, all of which should be recognisable from foregoing chapters. Thus there are *labels* that help you keep track of something by creating a vivid name for it, as I have tried to do with the 'dark self' idea. There are *examples*, such as the cameo from Porter on logging, to concretise otherwise rather abstract considerations. *Analogies and metaphors*, such as that of sustainable development as a set of lead spanners, play a similar but more generative and open-ended role, while *staging* builds up frameworks for getting at issues as a whole, rather than piecemeal – here, perhaps, my four slogans from the Prologue, aiming to get the focus onto tragedy and the present as against future-oriented meliorism, are doing this kind of work. (So I can in fact claim that the whole book has been engaged in offering a toolkit with practicality ultimately in view, whatever the appearances hitherto.) Finally, in this list there are *intuition pumps* themselves, which is Dennett's term for thought-experiments (thinking about change in the melting ice, for instance, as a means to clarifying our concept of the self that registers change).

The image of an 'intuition pump' goes nicely with the idea of a *toolkit*. But the metaphor has less helpful aspects too. A pump, simple or more complicated, inevitably suggests a process of extraction (as of shipped water from a boat, or oil from a well), applied to produce something that we know to be down there, waiting to be made available. For that reason I myself prefer the term 'intuition-*prompt*'. As conceptual tools, such prompts

are of the simplest. They are perhaps most like those inverted rubber suc-
tion caps on sticks that plumbers use to free up drains.[3] They are, precisely,
intuition-*unblockers*. We can get self-blocked or cramped into a particular
way of thinking; and this can go very deep, so that certain things seem to
us just utterly obvious, or such as no decent person would ever dream of
questioning, or on the other hand unimaginable or even impossible. Then
the process of unblocking will be the revealing of those axiomatic assumptions *as* assumptions, open at the very least to interrogation, and will allow
freer flow to intuitions about possible alternatives.

Interrogating our assumptions might seem to be something that can
only be done on the basis of other assumptions. But the most revealing such interrogation occurs when we can open up our awareness to
what comes from beyond assumption – that is, from behind the 'I' that
assumes. Unblocking can take us into the territory that I have assigned
to the dark self, the domain of what tacitly configures what we know or
feel. This is philosophically important in itself: the knowledge-framing
activity of the ego-self working its sufficient reasons can never encom-
pass or even reach to all of the inhabited sense that we make. But it is
especially important for our enterprise here. Clearing the channels for the
making of such inhabited sense is crucial to the kind of natural responsibility out of which, I have argued, any hopeful enterprise of retrieval
must come.

It is a corollary of this that, from the aspects of retrieval which I touch
on below, ways of similarly unblocking our thought should follow for all
those aspects that I have no space to discuss. I offer in this chapter to free up
our thinking about the key retrieval issues of resilience planning, education,
political legitimacy and internationalism – in part because these issues are
fundamental, in part also because they are those on which I have worked as
an environmental philosopher and have thoughts to share. The intuitions
potentially freed up have implications in turn for a wide range of other
matters – the arts, capitalism, national defence, immigration, international
cooperation, the internet, religion, the role of science and technology, the
nature of work – that I must leave the interested reader to identify and
explore. Within the constraints of such a book as this, I can only suggest
what some important kinds of conceptual unblocking might yield, and then
hand over the toolbox.

What I do have room to say, however, may well be found to go further
in several directions than many will initially like. Let us try how these tools
feel in use.

Community resilience: preparing for the unexpected

Resilience, as I have noted earlier in the book, must be a central concept for retrieval, which must depend on the capacity of communities to bring something like civilisation out on the other side of the *unpredictable* stresses and dangers to which climate change will necessarily expose us.

Rob Hopkins in *The Transition Handbook* defines the idea of resilience more specifically thus: 'the capacity of a system to absorb disturbance and reorganise while undergoing change, so as still to retain essentially the same function, structure, identity and feedbacks'.[4] (It is noteworthy that this is almost word for word the same definition that is used, or at any rate cited, in the UK Government's 2012 National Adaptation Programme.[5]) Hopkins glosses it as referring 'in the context of communities and settlements ... to their ability not to collapse at first sight of oil or food shortages, and to their ability to respond with adaptability to disturbance'. He recalls the notorious lorry-drivers' rebellion over the fuel tax escalator back in the year 2000 as an illustration of the current *non*-resilience of Britain as an economy and society. That crisis, in which the interruption of food supplies through the breakdown of a thoroughly centralised distribution system was recognised as briefly imminent, revealed our current version of 'civilisation' as 'only three meals deep'. The distribution system, like so much else in our current arrangements, was revealed as 'too big to fail' – that is, too big to be deliberately allowed to fail, which of course is what the lorry-drivers' collective blackmail relied on.

Retrieving resilience means taking the 'too big to fail' tendency out of all our systems. The components of resilience as Hopkins identifies them are *diversity*, *modularity* and *tightness of feedbacks* (what we might alternatively call *informational integration*) within a system. Respectively, these refer to:

- the number and variety of system elements;
- the extent to which these elements are not vulnerable to a 'domino effect', but can survive (by effectively self-reorganising) a shock that takes out some of their number; and
- the ability of parts of the system to register and respond to impacts on other parts, that is of the system as a whole to use internal information flows effectively in the process of self-reorganisation.

In its account of what will need to be done, locality by locality, to recreate these features in communities from which they have been progressively stripped over the past century by the centralising and alienating processes of an oil-based economy, Hopkins' *Handbook* seems to me second to nothing of its kind that has recently been produced. There is of course lots that is of similar practical interest going on outside the Transition network as such[6] but Hopkins' framework articulates especially clearly the impetus common to most of this activity.

Its leading themes for the transition that we must all now get ready to make are 'energy descent' (preparing to do with less centralised output, rather than relying on the implausible substitution of renewables for all fossil fuels at something near to current consumption levels); regenerating the local economy, in particular as regards food production (replacing 'food miles' with 'food feet', and similarly for much else of what we need); and rebuilding the networks and institutions of real human connectedness within which communities will have to operate to achieve and maintain these recovered local strengths as the globalised systems surrounding them start to unravel. These have of course been familiar themes of green thinking for as long as there has been a green movement, but they have latterly been much overlaid by the mainstreaming of that movement as 'sustainable development', the fantasy that we might continue a glib, facile and destructive way of life by relying on a combination of conservation management and technological smartness. The coming coincidence of Peak Oil with intensifying climate change would have had to mean the end of that fantasy, even though we hadn't recognised it as in itself humanly impoverished to the point of viciousness. The Transition movement that Hopkins has largely inspired is both a timely reassertion and a lively, congenial framing of what the green movement had kept in its heart all along. As can be seen from the rapid extension of Transition since its beginnings only a few years ago, it is also being found increasingly persuasive among people for whom the realities of our situation are becoming more and more unignorable.

For all that, I have to insist, it badly needs supplementing or informing out of the kind of thinking that we have been doing up to this point.

The easiest way to see this is to flag up a fundamental problem with resilience in this context, a problem which we can now turn to our first intuition-prompt in order to bring out clearly.

'Preparing for the unexpected' works as such a prompt because it is an injunction that is superficially easy to take as only superficially paradoxical. The superficial paradox is very cheaply obtained: if we are preparing for the unexpected, it can't be unexpected, because we were preparing for it. Evidently, the point of this small jolt of smart-arsery is to reinforce the (apparently) perfectly sensible underlying message that we need to be ready for anything, and in particular to be ready for what actually happens in any given case to turn out differently from what we had been anticipating would happen.

But if we persist in bringing the injunction's rhetoric of inversion to bear on this apparently straightforward second level of meaning, we can see that there is a much deeper paradox here. If being prepared means being ready for literally anything (physically possible) to happen, then all sense drains out of the idea of preparation. To be prepared for *anything* is actually to be prepared for nothing. (How could I be 'prepared', now, for the floor of this perfectly ordinary office building suddenly to give way, *and* for having a coronary before I get to the end of this line of type, *and* for the imminent phone call announcing my £10m legacy from an entirely unknown relative, *and* for the fire alarm to go off because of a real fire this time and not just another malfunction or student prank, *and* ...?) As a matter of logic, it would seem, expectation, and still more, *preparation*, must involve at least some focussed anticipation, and couldn't describe an all-round readiness for anything at all to happen even if such a state were psychologically imaginable. We can only be 'prepared for anything' when that means 'anything *likely* to happen'.

But then, here's the rub: any focussed anticipation of what we judge likely, equally as a matter of logic, must concentrate our attention on what is thereby prioritised as focal, and thus leave us to just that extent *unprepared* for anything different. So either way, being 'prepared for anything' seems to mean: *not* being.

The intuition which this deeper paradox prompts is, I suggest, that the process of building resilience is liable to be exposed in an analogous fashion. The National Adaptation Programme, for instance, affirms robustly that 'through good risk management, organisations can become more resilient'.[7] And it is certainly true that accurate estimation of the likelihood of something's happening can contribute to our developing the ability to respond to and hopefully survive *that* anticipated impact, if it does indeed materialise. But implicitly or explicitly quantifying risk in this way can also *reduce* resilience: it can focus anticipation and preparation in particular directions, and so generate increased path-dependent vulnerability to something *unanticipated* happening. This is all the more a danger, to the extent that we are moving in a domain of the unpredictable and in various ways indeterminate, as with environmental and climate futures we always are. In such a domain there is all the more danger that, as well as the consequences of simple ignorance, the quantified prioritising done in 'risk analysis' will actually reflect what we *want* to think probable or possible, or (still more insidiously) it will tacitly frame out what we just don't want to think about at all, and so the chances that what actually materialises will come on us unexpectedly are increased *pari passu*.

It is worth at this point rehearsing with more deliberate reference to our resilience-building requirements the grounds that we have for taking ourselves to be confronting what might well be called billowing uncertainty.

As we noted in Chapter 7, climate-change and environmental futures are not *completely* unspecifiable – what we must recognise to be coming isn't just random. These futures do, however, involve very significant uncertainty over, in the first place, which among the range of possible impacts are to be anticipated. The spectrum of possibility here stretches over the following (by no means an exhaustive list):

- changes in energy sources and availability (whether through oil-supply or carbon emission constraints);
- weather-related effects of climate shift or instability, including physical effects (such as sea-level rise, or heavy rain combined with diminished absorption capacities leading to flooding), ecological effects (changing growth patterns, increasing prevalence of pests, diseases and so on), and effects on human physical health (including heat stress, hypothermia and disease patterns more generally);
- food supply impacts of all the above;
- impacts on the habitability of different areas, and other land-use impacts (especially for agriculture and horticulture);
- employment and related economic consequences of any or all of the above;
- changes in social arrangements and patterns of behaviour and in institutions flowing from any or all of the above;
- shifts in external as well as internal sources of stress to which communities and individuals must respond, including immigration, refugee pressure, resource wars, related terrorism and security consequences; and finally
- psychological effects, cumulatively, of all the above (what Sally Weintrobe calls 'environmental neurosis'[8] and Hopkins, with a more peak-oil focus, 'post-petroleum stress disorder'), including depression, despair, overspill anger, and forms of compensatory fanaticism both political and religious.

Right across this spectrum, and in relation to each possible factor, uncertainty also attaches to the *scale* of impact in each kind (from small shift to upheaval), the *timing* of impacts (from short-term to distant future, several centuries out), their *interconnection* – how the above factors will each affect the others, from individual impacts all the way to the overall configuration – and their *significance*, in terms of our reactions to them and how these reactions in turn reinforce or ameliorate the impacts indicated (that is, become themselves second-order impacts, subject then to all the above kinds of variation).

It is thus very evident that uncertainty as to the scale, timing, interconnections and likely significance of potentially all the impacts across this full range amounts to effective indeterminacy within the arena of climate change consequences. To know what we know when we have stopped pretending – that is, that what is inescapably coming is going to be somewhere on the range between very severe and catastrophic – is not to know in terms of which combinations of factors those conditions are actually going to manifest themselves. Indeed, the extent of our inevitable ignorance here is an important part of what we *mean* by 'severe to catastrophic'. In this context, building on-the-ground resilience (which we must do, because it is all we *can* do) must involve – otherwise we should never get started – building capacities to pursue a certain identified pathway through all the above. An example is Hopkins' local Energy Descent Action Plans or EDAPs. The EDAP template as he characterises it is explicitly path-dependent: 'An EDAP sets out a vision of a powered-down, resilient, re-localised future and then backcasts, in a series of practical steps, creating a map for getting from here to there'.[9] But creating genuine resilience also means designing into our systems 'real options' (for instance, physically realised switching capacity as between different renewable energy sources[10]) to keep open the possibility of having to shift onto other plausible pathways, sometimes at short notice. As a corollary, it requires us to build in a review and revision function, to ensure that scenarios, options and plans are re-crafted ongoingly as necessary in response to emergent conditions (as far as it is possible to do so, because of course options are always closing as well as opening as we move on). Moreover, making this function effective involves maintaining as far as possible not just alertness to the current plan (tracking milestones, indicators and measures) but the very best we can achieve by way of what might be called 'full-spectrum alertness', in order to inform this ongoing review function with earliest-possible awareness of all relevant *unanticipated* factors that may be emerging to affect scenario plausibility, option status, and therefore the need for new arrangements ad hoc.

The crucial thing in all this is that under ramifying uncertainties of the range, scale and volatility outlined above, 'full-spectrum alertness', the fourth requirement just noted, is going to be absolutely vital to genuine resilience – but it calls for a kind of non-directed attention that option- and scenario-based planning will inevitably tend to be channelling in particular already-identified directions instead. So there is a clear and potentially disabling tension between the first two and the second two requirements for building on-the-ground resilience. Given a broad spectrum of fairly open possibilities, organising for real action always involves taking a bet on *reduced* full-spectrum alertness. This process we may reasonably describe as one that under significant uncertainty decreases our flexibility of response in the very act of trying to increase it.

The issue here, of course, is that of Donald Rumsfeldt's notorious 'unknown unknowns'[11] – the upcoming shocks that escape getting factored into risk quantification because they are off our risk–assessment radar altogether, so that we don't even register that we don't know their probability of happening. For example, suppose we build up local food supply chains, and then they are hit by a completely unpredictable virus from some imported product. Or suppose we invest heavily in solar panels to feed a local grid and provide 50 per cent of local supply, and then a freak storm of entirely unprecedented violence rips them off all the roofs … Building capacity to absorb disturbance and system shock involves anticipating likelihoods among potential sources of that shock, and in a situation of significant uncertainty this is likely to veil from us 'left-field' possibilities of these kinds. The danger that shock will come from unknown unknowns is indeed increased in a situation where we think we have our bases covered in respect of the known unknowns, because our plans and preparations for building path-dependent resilience actually *create* unknown unknowns just insofar as they focus attention on the possibilities for which we know we have to condition. How do we prepare for *those* shocks? – given that, as outlined in the previous chapter, we have good reason on general grounds to think that many of them will be in the offing, and therefore little reason to think of our systems as adequately resilient unless they *will* somehow be able to deal with them.

Confronting catastrophe: 'wild planning'

If we are confronting potential catastrophe coming at us out of the unknown, the only possible appeal here is to a notion of responsiveness without control, or poised spontaneity, which we might call 'wild planning', that is at first blush at least as paradoxical as preparing for the unexpected. Wild planning, preparation that is not under our control, is that intuitive readying for whatever might come at one, that intense non-specific sensitivity to its impinging environment, which a wild creature must constantly deploy in order to survive: a matter not of identifying what is likely, but of living in and from the permanent possibility of the unidentified and indeed unidentifiable. In human beings, this must rest on a capacity and readiness to deploy alertnesses that we don't know we have, in support of those of which we are aware.

What is involved here can be suggested by another intuition-prompt. 'How do you know what's going to happen until it happens?' is a familiar enough wry question which we can easily imagine addressed to someone who seems too breezily sanguine about his or her ability to know what is approaching from over the horizon. In this usage, it expresses, apparently, a kind of folk-recognition that knowledge of the future is impossible. Until whatever is in question happens, we can only be guessing or estimating probabilities, which falls short of knowledge. But once it has happened, our knowledge of it is no longer knowledge of the future: *knowledge* of the future always comes too late for itself, as it were. The vulnerability exposed here is of a kind which can, on reflection, be seen to attend all our knowledge of the future as a matter of logic. As we noted from Hume's Paradox in the previous chapter, we can never have rationally adequate justification of any claim about what will happen, since all justification based on past or present experience must simply assume its own continuing justifiability in relation to the future; and therefore no such claim can count as predictive knowledge, which must imply adequate justification. When we do arrive at adequate justification based on present experience, however, the claim has already ceased to be a prediction.

All that is undeniable, but what is also plain is that nothing capable of envisaging its own future, about which that was the whole truth, could possibly have survived. Creatures that can register the world as including a dimension of futurity have to be able *reliably to anticipate* that future on at least the large majority of occasions, in order to survive – since consciousness of that kind brings with it also the capacity for self-delusion into a false sense of security, and unless a creature is able *routinely* to correct the associated tendencies, it will not be fit for any environment in which it finds itself.

Human beings are no exception. We 'reliably anticipate' by reflexively conscious prediction, which ordinarily involves assigning probabilities, and in the standard case where some kind of action has to be based on the anticipation, we rely on this assignment in order to invest preparatory effort proportionately. Suppose we had assigned to outcome A a probability of 0.6, to outcome B 0.3 and to outcome C 0.1. Rationally, we would not then invest all our effort in preparing for A, but equally we wouldn't invest all that much in preparing for C on the off-chance. Thus, for a simple (non-quantified) example, I might judge that it will *quite probably* rain, because those dark clouds to the west *most likely* mean rain and they will be driven in this direction unless the wind changes, which it *shows no signs* of doing: and I might therefore plan to do something requiring a significant time-commitment indoors, but not something that I absolutely couldn't put on hold in case it doesn't in the event rain and I can get out into the garden.

Here my judgement of the probability of its raining expresses my degree of confidence in the way the evidence of the clouds and of the wind combines – the rain is only *quite* probable, because I *could* be wrong about either.

But now it might seem that a question should arise as to how probable it is that I have got my judgements of the evidential factors right. That would be to ask, for instance, how probable it is, and on what grounds, that the given darkness of the clouds *does* on this occasion portend rain – and if I take that to be highly probable on the grounds that all such clouds that I can recall have yielded rain, how probable it is that my memory isn't playing me false here? Clearly a regress is in prospect, which I can only stop by taking some level of evidential grounding as non-probabilistically given. But I cannot do that on the basis of what I *know* about the coming rain, since what I *know* has already been summed up in my judging that rain is quite probable. Thus, not everything I know about the future can be probabilistically warranted if I am to have probabilistic knowledge at all.

By the same token, my practical confidence in assigning a probability of 0.6 to A is expressed in my giving that level of emphasis to A in preparatory action, which means that I have to take such an assignment of probability as *robust enough* for such action. But again, I can't do that on the basis of what I *know*, since all that I know has already been committed, as it were, in the assignment of probability. We must always be in possession of more grounds for the assignment of an operational probability than are ever explicitly committed *in* that assignment.

It follows that *all* planning must be fundamentally 'wild', in the sense that it must tacitly rely at some level on our having more grounds than we can know ourselves to have about the probabilities in question, and so on our ceding authority, just so far, to a warrant that must shape our plans from the life in us that lies beyond our conscious awareness and cognitive control. The application of this to planning for resilience in face of serious climate jeopardy should be clear. The greater the uncertainty under which we are predicting and planning, and the more potentially destructive and even catastrophic the unknowns could prove, the greater our overt reliance on the element of wildness in all planning has to be.

A suggestive analogy (though no doubt an initially surprising one in this context) is that of *battle-readiness*. A soldier going into battle confronts an individually catastrophic possibility – his own sudden and violent death – under conditions of almost complete unpredictability. He cannot do this as an ego-self, since 'my life is for me' would be a paralysing awareness at that juncture and must be transcended for him to act. Armies have training and drilling methods to ensure that this happens, making disciplined submission to the goal-directed activity of the relevant collective – unit, platoon or other formation – a kind of second nature even under these conditions. The component corresponding to wildness here is the subsumption of individual awareness of dangers and opportunities into the multiple, diverse and de-centred awareness distributed across the collective entity of which the individual soldier feels himself an integral part. The soldier is able to go into battle resolutely because he goes not only with the multiple eyes and ears, but also with the common being and purpose and the (comparative) indestructibility of his unit, just as a wild creature can be protected not just by its own alertness but by a shared alertness distributed through the group or flock in which it belongs. Achieving goals under conditions of battlefield danger needs a collective form of survival knowledge in which each individual participates, and on which he relies for his chances of individual survival, but which he cannot know himself, *as* a self, to have. Another way of putting this is to say that the soldier goes into battle from his dark self, from that in him which is 'behind the I', which if he is well trained will include that quasi-instinctive reliance.

To propose such an analogy might seem only to emphasise how far these conditions of 'wild preparedness' are from anything that we can yet recognise in the first, hesitant processes of retrieval. But that it is possible for a whole community to experience a transmuted form of battle-readiness is historically attested: Churchill, for instance, writes of 1940 as a period when 'there was a white glow, overpowering, sublime, which ran through our Island from end to end', and also as 'a time when it was equally good to live or die'[12] – and even when allowance is made for his personal involvement and romantic nationalism, *something* of such ultimate buoyancy is surely indispensable from the kind of collective mood which we are again going to need. Jimmy Carter spoke more truly than he knew when (hijacking a phrase from William James) he called an earlier glimpse of what we are now facing 'the moral equivalent of war'.[13] Moreover, once we admit that we are in that order of jeopardy, we must also recognise that we are building resilience not for a transition from unsustainable to sustainable comfort and well-being ('a better world for all those who come after us'), but for surviving an era of hazard and deterioration that will leave whatever can be retrieved from it scarred, shaken and partially devastated. The war analogy here is very far from gratuitous. Anything that helps us to think of where we are headed not in terms of *News from Nowhere* (which often seems to be the back-of-the-mind ideal here[14]), but of something much more like Europe in 1945, is to be welcomed.

But the question is then, how do we bring that kind of recognition to bear on our present case – on the happy-Transition-towns, Garden-City and farmers'-market models of retrieval that are all we presently have to work with?

We suffer in this context from the fact that currently Transition protagonists, and more generally environmental activists, tend to come preponderantly from the left-liberal end of the Western political spectrum. There are historical explanations for this fact, linking it to progressivism and the recent hegemony of sustainability discourse, but I have no space to go into them: we must focus here on the consequences. Such people incline to be suspicious on principle of anything that downplays the role of the ego-self. (A liberal outlook, indeed, might fairly be characterised as the ego-self in its political aspect: my life is for me, yours is for you and we negotiate our common interests on that basis.) They are thus predisposed to disregard, where they are not actively hostile towards, the kinds of association in which wild preparedness could flourish: those constituted by something that the idea of the dark self suggests that we could call 'dark community'. Dark community is held together by local or patriotic feeling, from beyond its members as individuals, in forms of unity that are not negotiated or conditional but borne on the currents of livingly coherent cultural tradition. For building real resilience, however, that kind of association is going to be indispensable. To confront dangerously unknown unknowns, we shall need to base our common action not exclusively in intellectual analysis and means–end rationality, but fundamentally in what the philosopher Roger Scruton has dubbed *oikophilia*.[15] This is love of the household, or more broadly of *home* – meaning, wherever we can think of ourselves as unconditionally belonging, such that it is an actual or imaginable object of one of the most basic forms of loyalty. (Among the main reasons why a bureaucratic multinational construction such as the European Union is so unconvincing and impermanent is that no one could imagine being seriously loyal to it – dying for it, for example.) For the kinds of reason that we have been considering in this section, localities and territorially defensible nation-states towards which such loyalty is strong, or could feasibly be strengthened, offer by far the best prospects for retrieval in the conditions that are coming.

But these requirements are a long way from being widely recognised as things stand. How then to proceed? The only thing to do here, I suggest, is to approach our job of making the dispositions required for on-the-ground retrieval in the spirit of what I have called existential resilience. The more we condition for retrieval and survival in acknowledgement of our tragic situation (including our inability to predict or control and the absence of guarantees), the more we will be opening ourselves to a kind of attention that corresponds as closely as we can yet come to the relevant kind of 'battle-readiness' – and the more, therefore, we shall be disposed to inhabit or reconstitute the kinds of association which can support that attention.

To the extent that we open ourselves to trust in our capacity to respond creatively to unknown unknowns, left-field possibilities, we will also find ourselves gravitating towards institutions of wild preparedness for facing potential catastrophe – towards forms of dark community that know more than we can know ourselves to know, and help us find more in ourselves than we can individually recognise to be there.

Tapping thus into our capacities for this order of resilience is what this whole book is really about. I have called it *existential* resilience because it is a way of resiliently decentred *being*, unsubverted as far as is humanly possible by the ego-self. In terms of getting through our individual lives, it is the ability to live intelligently from our human wholeness while engaging in the subject-centred cognitions and commitments that conscious active life inevitably involves. It is the ability to live in practical optionality from human wildness as far as we can.

In terms of what we need for any hopeful retrieval in our current plight, existential resilience is the addressing ourselves in that spirit to the best collective preparatory action for on-the-ground resilience that we know how to take. We must try to shape the best practical outcomes we can envision by our best lights, while recognising that the upshot of this work is always out of our control, including our control of what *tries in us*. From a religious perspective, this might be described as doing our human best, while humbly leaving the event in the hands of God. The non-religious (or at least, non-denominational) form of that humility is knowing that we are always acting from more than we can know and that we do not belong to ourselves.

The relation of existential resilience to practical retrieval thus looks both complex and cyclical. There are alienated ways of life, such as the current urban-megalopolitan norm, which could not plausibly foster such resilience because they are so adverse to natural responsibility and so liable to foreground exclusively the 'for-me' of the ego-self. The further we break away from such life-modes towards retrieving genuinely robust local economic and social arrangements, the closer the contact into which we are likely to be brought once again with non-human natural systems, and the richer, correspondingly, the soil in which existential resilience might re-grow. But the building of genuine economic and social resilience will depend (as we have seen) on our having *already in place* some measure of existential resilience, some capacity for the life-intelligence that must underpin proactive planning under serious uncertainty. So the key question is how to break into the cycle. How can we condition for existential resilience as part of any actual practical process of local community or inter-communal work towards retrieval?

That is a question that in a book like this I can hope to do no more than open up. For one thing, accessing our wholeness in action certainly depends on what we might call full-spectrum *intelligence*, the developed human manifestation of the full-spectrum alertness deployed by other creatures. This issue is key for the closely related matter of education for retrieval (or, for resilience), about which the next section seeks in turn to prompt some intuitions.

Education in transition: knowledge for its own sake, or for the sake of retrieval?

The *Transition Handbook* offers what is in effect a *blueprint for resilience*. That is a very far from sufficient requirement, as we have now abundantly seen, but it is a necessary one. When I said in Chapter 7 that there are going to be no blueprints, I meant of course no detailed plans adequate to doing the *whole* job that needs to be done. The point of a blueprint after all is to *be* sufficient – a partial blueprint, or one with bits that clearly need supplementing, is not really a blueprint at all, since it is of the essence of a technical or engineering design that it be complete. (The one that says 'OK, improvise this final bit of the bridge where it gets towards the other bank' will not only not get selected for implementation, but will be unreliable in what it *does* specify.) As we have just seen, however, it is of the essence of a genuinely resilient plan under uncertainty that it *not* be 'complete' in this way, and regarding such plans as blueprints – a scientistic, or strictly *technicistic*, aberration – is a real problem for retrieving resilience. We have to see planning under real uncertainty as something *always upheld by open-ended creative intelligence*. But where in preparing for retrieval do we condition for developing that capacity?

In one chapter of the *Handbook*, Hopkins imagines himself looking back from the year 2030 on various aspects of a society that has by then successfully 'transitioned'. About education he has this to say:

> Education in 2008 [the date of the *Handbook*'s publication] was woefully inadequate, given the scale of the Transition to come … young people leaving school were unprepared for the more practical demands that the emerging powered-down world made of them; their school years had left them unable to build, cook, garden or repair … From primary school level upwards, gardening, cooking and woodwork skills have become a core part of the programme for the first time since the 1950s. School grounds have been transformed into intensive gardens … By secondary school age, students now learn construction, as well as

creating, installing and maintaining renewable energy systems, alongside social skills like conflict resolution and community leadership. For
adults, Colleges of the Great Reskilling are now central to most towns,
offering a variety of courses in a wide range of practical sustainability
skills for the public as well as retraining for professionals ... By 2018
many of the larger comprehensive schools and universities were no
longer able to attract their intakes from large areas ... they diversified,
and are now also homes to incubator units for new businesses, with
skilled craftspeople having their workshops and offering apprentice-
ships onsite.[16]

Both the explicit and the implied critique of present education, with its
obsessive scholasticising of every study and its demeaning of practical skills
in the service of frequently meaningless examinability on paper, are very
well taken – and the shift of emphasis envisaged is clearly going to be vitally
important for retrieval. What this vision doesn't touch on, however, are the
questions of what balance to strike between acquiring these kinds of useful
skill for resilience and the continuing demands of wider cultivation – of
initiating students into historical awareness, artistic and literary sensibility,
an appreciation of science not driven by technological challenges, and so
on. It would be an unfair simplification to say that for Hopkins all we really
need to understand about the human condition is Peak Oil, the facts of
climate change and the application of permaculture principles and practices
to local economies. But he conveys in this significant passage little sense of
recognising that an overwhelming practical focus on re-skilling people for
Transition would involve as much of a travesty of real education as does the
focus on skilling them, as we do now, for servicing the globally networked
petroleum-based economy.

A link back to the paradoxes of resilience just discussed should be evident
here. Focus, while in one sense clarifying what is focussed on, inevitably
also *veils* what is out of focus, and veils from us that it is veiling it. Thus
not only are we rendered vulnerable to shocks from within or beyond this
penumbra, but also – and crucially, when education is at issue – at least
some of what is focal is in another sense itself obscured, since our awareness
of its broader human, natural or technological *context* is reduced.

We shall need, in the interests of this awareness, to go on being able to
read Shakespeare and Descartes and Newton and Darwin (to name merely
some representative names), and to experience the exhilarations of mathe-
matical insight and of scientific curiosity, as well as being able to cook, mend,
build and garden, however stringent the demands of transition and however
drastic the collapse to which it responds. Otherwise, whatever we do manage

to retrieve will no longer be civilisation, nor indeed recognisably a human future at all. But for all that, some demanding trade-offs are evidently looming. There will certainly have to be some serious triage applied in the process of moving from our current educational agenda to one plausibly adapted to retrieval. Our need for useful knowledge and skills of the kinds that Hopkins identifies is rapidly increasing, and correspondingly our latitude for the pursuit of knowledge 'for its own sake' must surely be diminishing, as crisis presses more and more sharply upon us while days remain firmly restricted to twenty-four hours each. How to exercise such triage?

That question points us towards our next intuition-prompt – in this case, an apparently enticing (since so often reproduced) dichotomy. The picture of education as providing its recipients with *either* useful knowledge *or* knowledge for its own sake is a very familiar one (a still-classic account in terms of higher education is Newman's in *The Idea of a University*, recently invoked yet once more in a discussion of very contemporary issues by Stefan Collini[17]). The dichotomy seems at first glance grounded in a genuine distinction. What we get from education must surely be either useful towards our ends (that is, knowledge of how to do the things we want to do, and of relevant facts), or an initiation into appreciating those ends themselves – into knowledge of things, facts, artefacts and human relations that, in coming to know them properly, we come to value intrinsically and not as means to other things. The distinction itself goes back to Aristotle, and appears exhaustive: that which is valuable must be so either in itself, or by reference to something else that is valuable in itself. What other options could there be?

When confronted with any dichotomy, however, it is always worth asking whether there is any third possibility to which we can jump. And here there evidently is. What the distinction leaves out is whatever we get from education that empowers us *to make sense of* some things as intrinsically valuable, and so to create for ourselves any ends we have.

This process of empowerment can't be valuable just in itself, since it is clearly *for* something. But that is not something in respect of which it is *useful*, either, since our sense-making power is what supplies us with criteria and warrant for calling anything useful. It can no more be useful, or not, than Parliamentary law-making can be legal or illegal. And yet, if education doesn't empower us to make sense in terms of intrinsic ends, it is clearly missing out something vital. So the dichotomy fails.

The intuition with which we are prompted by its failure is that there must go on, within the broad field of education, activity that is best thought of not as the transmission of recognised knowledge and value nor as training in specific skills, but as a fostering and empowering of creative intelligence.

The intelligence that creates value by underpinning and warranting all evaluative judgement cannot be valuable in itself, since that would be to have always been anticipated by itself, nor can it be valuable for the sake of other values, which would be to be always too late for its own task. What gives all 'sakes' their sense cannot exist either for its own sake or for that of something else. But it is what we supremely need for our lives as sense-makers.

It is of the nature of such intelligence not to take any framework for its exercise as given – whether this be the needs of the globalised economy, or of any particular vision of transition or retrieval – since whether any framework is *rightly* given is always the responsibility of intelligent sense-making to decide. By the same token, educating to foster creative-evaluative intelligence is inducting people into an activity that must take responsibility for itself, because it must always be deployed *without any criterion*. Intelligence takes responsibility for itself by seeking to realise itself as its own best – as the fullest exercise of the capacities inherent in the idea of intelligence itself. It aspires, that is, to the completest alertness, ratiocinative acuity, articulacy, sensibility, delicacy, integratedness and analogical energy of which we are capable – and in a way for which there cannot in the nature of things be any standard except its own self-realisation. (We should always have to ask whether any proposed standard was an intelligent one, or being used intelligently.)

Evidently, we are dealing here with the 'full-spectrum intelligence' which I was invoking in the previous section as a crucial element in existential resilience. We might call it, in its full human perspective, *life-intelligence*. Equally evidently, it cannot ever be thought of as a capacity under our control, and for just the same kind of reason as we have been canvassing: if we controlled our uses of life-intelligence, we should have to do so life-intelligently otherwise it would not be the use of intelligence that we were controlling. Life-intelligence can only be appealed to as *its own criterion* of control. Intelligence thus criterionless must come out of the whole organic-conscious self, since that is the only source from which it can draw on capacities and recognitions both fundamental to our lives and necessarily beyond our control, but which we nevertheless have no option but to *trust with* our lives. Creative sense-making cannot be by the ego-self. That which shapes all the sense I can make would be denatured if we thought of it as in any sense *for me*. Significance can't be me-centred, because if I judged what mattered in that mode, its mattering would not *orient* me in the world. (A world centred on me cannot be one in which I find my way – that would be like every map representing only what I could see from where I was at the moment standing.) Evaluating (sense-making) intelligence must come from beyond the *me*, reflecting – or realising for us – our being de-centredly in the world.

How does all this, arising from our intuition-prompt, help towards education for retrieval? It shows us why we shouldn't really be thinking of the fostering of this kind of life-intelligence as education *for* retrieval at all, but rather of the education that empowers it *as* retrieval. Not as the whole of it, obviously, but as the essence without which it won't be retrieval of our *humanitas*, nor the re-establishment of any genuine resilience. And by the same token, what we offer students at any level won't be real education unless it is conducted within a context of genuine retrieval, lived from the dark self and informed by natural responsibility. It is the free flow of an intelligence that is always greater than we know ourselves to have, or can consciously bring to bear, that constitutes the heart of existential resilience – without which measures for the recovery of economic and social resilience will always leave us still vulnerable to the shock of the emergent.

Of course, such recovered life-intelligence doesn't just *happen*, even in the context of deliberate socio-economic retrieval – it has to be actively fostered, though it can't be taught, only shown forth in divers ways. I cannot do more here than gesture at how such fostering might be approached in a community bent on retrieval, but a much greater emphasis on the arts and humanities in non-vocational education at all levels is an absolutely key requirement – as is the corresponding insistence on the non-vocational core of education even when real practical needs are pressing. (It is indeed when they are most pressing that the vitality of this core most matters.) All real art, properly responded to, calls forth our attending dark self in its wholeness, and thus images back to us our wildness, just as does the natural world in the way we were considering in Chapter 6. Nature makes a simpler and more spontaneous sense of our wildness, the flesh and blood and growing fibres of the natural world speaking to ours and resonating with them at a deeper level than the conscious attentiveness that art requires. But in art in all its variety, as Dilthey observed, '*life discloses itself* at a depth inaccessible to observation, reflection and theory',[18] and does so through all the multiplicities of human life that art takes for its subject matter. An education informed by an active, developing and engaged familiarity with even one or two modes of this variety (the significant literature of their own language is probably for most students the most accessible) can initiate unsurpassably into life-intelligence.

Equally clearly, the arts cannot stand on their own for such an education. If art gives us the exercise of such intelligence, language gives us one of its major means, history its contexts, philosophy its most critical conceptual junctures. All this and more needs to be rewoven into a curriculum also vigorously enriched with the renewed practical affordances indicated by Hopkins, if we are to have any hope of retrieving genuinely resilient communities.

We come back here to the point, on which we have already touched several times, that hopeful retrieval starting from where we are now cannot just be of the external conditions of ongoing life. It is not simply a viable pattern of providing for ourselves despite increasing climate and environmental dislocation that we need to reconstruct, but a habitable *humanitas*. Education as currently conducted in the advanced societies of the West and North has largely lost sight of its obligation to initiate young people into the living comprehension of human nature and possibility that this implies. That is not merely a matter of its grotesque failure to provide them with an understanding of what is really at stake in respect of our environmental situation, although such failure is a major symptom. The essential disease is the loss of any sense of self beyond ego-self in perhaps the most crucial arena where that sense needs to be fostered. Across the board, our scientised, technicised and generally instrumentalised education now characteristically fails to suggest the human depths and realities, without a sense of which no real existential resilience can be developed. It therefore sells any mature idea of human possibility drastically short, and leaves the large majority of students disinherited and helpless in face of precisely the kind of tragic situation that this book has been trying to explore. We shall not retrieve that situation without also retrieving ourselves.

But such retrieval, educationally vital as it is, has also an inescapable political dimension, to which we must now turn.

A vanguard movement? – 'Says who?'

The political philosopher John Barry, himself actively involved in the Transition movement, calls its committed activists 'pioneers', and quotes with approval a claim that well expresses the pioneering spirit: 'the only way to give the next generation a decent shot at life is for those of us who care most about them to take things into our own hands and prepare for the changes ahead'.[19] But he also, in the same breath (or at any rate, in an immediately contiguous footnote) disavows what might seem to be a clear implication of 'taking things into our own hands' – that Transition is a 'vanguard' movement. There is an important tension revealed here that we shall certainly encounter in practice as we start to take attempts at retrieval beyond locally-based economic initiatives, and beyond educational concerns that might easily be misidentified as specialist, into the wider political context. Even the most basic conceptual toolkit must therefore include tools for thinking about this tension.

It is true that Barry qualifies 'vanguard', in that disavowal, by adding 'in the classic Marxist-Leninist sense'. But this caveat is not enough to remove the tension. No doubt we should look in vain among Transition pioneers for professional revolutionaries operating in armed secrecy to evade the attentions of the political police (though recent revelations about the UK and US security agencies might make one wonder about that last bit). Consider, however:

> The Communists ... are, on the one hand, practically the most advanced and resolute section of the working-class parties of every country, that section which pushes forward all others; on the other hand, theoretically, they have over the great mass of the proletariat the advantage of clearly understanding the line of march, the condition, and the ultimate general results of the proletarian movement.[20]

Substitute, in that passage from the *Communist Manifesto*, 'Transition activists' for 'Communists' and 'environmentally concerned people' for 'proletarian movement' (and similar), and you have perfectly characterised the nature and grounds of the pioneering role that those committed to retrieval on the basis of having stopped pretending must now assume. The key issue raised by acknowledging this surely inevitable vanguard role is: how should that kind of vital energy and impetus – exercised in a networked but essentially independent way in a variety of local contexts – relate to the political processes of the democratic state?

The Transition movement itself has been criticised as anti-politics, and the leading Green philosopher-politician Rupert Read has taken it to task for encouraging its devotees to think that 'Peak Oil renders ordinary politics irrelevant'.[21] But its real tactic so far has surely been to *bracket* ordinary electoral politics, at any rate at the national level. The knowledge that this level matters unignorably has been deliberately postponed to the more pressing need, as matters presently stand, to get a good number and variety of real demonstration projects up and running. As Hopkins himself says in a response to Read, however:

> Transition Initiatives are seen as one of a hierarchy of approaches that will be required to get us through the twin crises of peak oil and climate change. We will need international action such as Contraction and Convergence, the Oil Depletion Protocol, strong international climate legislation and a moratorium on biodiesel production. We will need national action such as strong climate legislation with realistic targets, a carbon rationing system such as Tradable Energy Quotas and a national food security strategy, and we will need more local solutions.[22]

We shall also (though such points tend not to get made from this particular quarter) need, for any viable retrieval, action by relevant nation-states to preserve internal order against forms of breakdown that Transition initiatives will not by themselves avert, to defend national borders against climate refugees impending in numbers sufficient to overwhelm all transition prospects if permitted entry, and to defend the security of 'transition space' against many other and less predictable threats in an increasingly destabilising world.

Retrieval, in other words, will have to be pursued politically in the nature of the case. Whatever different emphases are laid on different aspects of the whole process for tactical reasons at various times, all these aspects will have to mesh together and collaborate if anything adequate is going to have any chance of being achieved. But then if we ask what is the tacit model for this collaboration, the words and actions of most participants and commentators so far suggest that it is something like this: groups initiate projects to retrieve resilience in specific local settings; these catch the enthusiasm and spur the imaginations of wider and wider circles of people; as this more widespread popular acquiescence is gained, and led by institutional opinion-formers in Parliament, corporations and the media who have already made the essential mind-shift, people begin to support the necessary enabling and defensive measures through the ballot-box – until eventually, the nation-state is signed up by due democratic process and its international actions and relations adjusted accordingly.

The key feature of this model of the process is the distinction that it embodies between initiation and legitimation. A group with a particular interest (even an interest so all-important as retrieving *humanitas* from disaster) is entitled to propose and energise and persuade, as well as to change its own practices within its own proper sphere of legal action, but it only acquires legitimate power to frame the institutions and practices of the whole community by gaining the endorsement of a democratic majority.

There are, however, two glaring defects attending this model in the present context, and it is well that we should confront them honestly.

The first is that we are where we are because democracy has failed – in relation to environmental issues, disastrously. *Democracy* of course is a hurrah-word (until one remembers Plato), but *mass-mediated soundbite-populism* perhaps evokes less knee-jerk approval, and that is now what we are really talking about. This media-based populism, to which universal suffrage combined with mass-communications technology have inevitably led, has simply been unable to recognise the green imperative. It is not just that, as Hopkins rather acidly notes, 'vote for me, and every year your consumption of energy, carbon-producing goods and services and travel will fall, but you'll be happier for it' is a very difficult message to sell in this kind of sound-bite market. The difficulty reflects the very deeply embedded features of modern mass-democracy that have helped to produce and confirm in place an unprecedentedly materialist and destructive world-society – one that in any other than material terms is helplessly impoverished, having no politically usable language *containing* any other terms. As a result, the translation into practice of the kinds of recognition necessary for retrieval – recognition not just of facts and of scientific authority, but of the tragic human depth of our situation as canvassed in this book – is incompatible with a system in which policy has to be expressible in a few minutes to an interviewer. But that in turn is the only way in which policy can be justified to a mass audience whose attention span has itself been drastically curtailed by habituation to the television and whose sense of entitlement to undiminished material well-being, as well as expressing the spirit of the age, is constantly pandered to by the popular press.

That is perhaps a hard saying, but it must not be shirked. For recent proof, if proof seems required, recall the US Gulf Coast *Deepwater Horizon* oil disaster in 2010, and President Obama's response: not to the oil coming ashore, which was maybe the best that could have been coordinated at short notice, but to the implications for fossil-fuel dependence. When Obama, probably the best Presidential intellect since Woodrow Wilson, spoke to the nation from the White House about these events, even he was reduced to misrepresenting the challenge, in covert appeal to the Al-Qaeda example, as something coming from *without*, 'an oil spill that is assaulting our shores and our citizens'.[23] It is surely obvious that no one who could say to the American way of life: 'Look, you really did have this coming – now let's face the implications and learn from them', could ever have been elected through the processes that we know about to the office from which it needed to be said. (The case of President Carter, who got elected in 1976 mainly by not being Nixon or Ford, and did little in office to match his initial green rhetoric, offers no real exception here.) But those processes simply represent mass democracy in its least qualified form.

The second glaring defect is a corollary of this first one. In practice, the situation in which we are most likely to find ourselves if retrieval is going to stand any chance at all, will be one where local resilience-building initiatives are becoming reasonably widespread, and where correspondingly there is growing recognition among at least some people at or near some important levers of corporate, institutional or political power at the national level, that those initiatives need to be promoted and empowered – which will mean matching their impetus by national and international action.

Meanwhile, there is likely to be a noticeable increase in popular endorsement, but still very far from a popular majority in support of such measures. A constituency for retrieval, that is to say, is likely to have emerged and at least tacitly coalesced, and to dispose of real opportunities to be effective, long before it could hope to have any significant use of those opportunities formally legitimised through the ballot box. And yet, while we cannot know what is coming when, we can be sure that it would be perilously easy to go on doing too little, too late. Is an effective constituency for retrieval then to refrain from precipitating state action as far as it can on its own initiative? That would seem grievously irresponsible. But on the other hand, *not* to refrain (to behave as a vanguard, in fact) would seem to be an act of presumptuous and illegitimate self-appointment.

I do not pretend to be able to predict, still less to advise on, how this deep but apparently unavoidable tension might be resolved in the actual politics of any existing or possible jurisdiction. Let me repeat that all I offer in this chapter are conceptual tools – ways of thinking about the concepts that knot together at the root of such difficulties. Here I believe that the bind in which we are caught *is* fundamentally a conceptual one, to do with the way in which we conceptualise legitimate political authority. As a step towards freeing ourselves from it, we may invoke another intuition-prompt. This is suggested by a different political philosopher, Jonathan Wolff, who opens his classic *Introduction to Political Philosophy* with the idea that there are really only two questions in his field: *Who gets what?* and *Says who?*[24] We can ignore the distributional issue for present purposes, and concentrate on that of authority. Where does political authority ultimately come from? How do we answer, in response to claims to exercise that power legitimately, the *Says who?* question?

The difficulty immediately revealed by this question is that, in response to the appeal to any criterion of legitimate authority, we seem perfectly able just to go on asking it. Suppose the appeal is to inheritance, for instance – rightful political authority is passed down from those who already have it. But who determines whether the criterion being invoked – the bloodline, say, or the choice of successors by the presently authoritative – is still appropriate to changing circumstances each time it is invoked? The answer in practice is always: those invoking it. But if conformity to the criterion *confers* authority, it cannot itself be authorised by those on whom it either has conferred it, or might confer it – any more than I can legitimately write a testimonial to my own truthfulness. Here the essential regress is sharply exposed – and clearly it is of the same order as is pointed to by Juvenal's ancient political question: *quis custodiet ipsos custodes?* – who guards the guards themselves?

Exactly the same problem arises if we say that scientific knowledge, or any other kind of empirically based expertise, confers authority. Such knowledge is always at least implicitly contested, and formally open-ended – and whence is conferred the authority to arbitrate particular specific contestations in favour of those claiming authority on the basis of what they take to be unchallenged knowledge? *Within* any community of empirical knowledge, authority is conferred in practice by consensus founded on repeatable experiment. But the issue here is authority extending beyond the consensus. Although empirical testing is in principle repeatable by anyone, expertise is always legitimated in practice within the community of experts through exchanges to which only experts can properly be party. *Says who?* – the experts. And says who that they can say? The answer can only be that they say so themselves, but here it seems as though that answer will not do.

It is this always-lurking regress that has helped to make democratic head-counting an apparently attractive source of political legitimacy. *Says who?* – says a majority (or in some cases, a decisive majority) of those relevantly consulted: end of story. And so, for many of the purposes of normal regulatory and distributional politics, it is. But of course it is not always the end of the story. As we have seen, there can arise on particular issues – as clearly there do arise on the issues of retrieval – very critical questions about who *is* relevantly to be consulted, and it cannot be simply *given* that the answer is always: every citizen in a mass-mediated populist democracy. But then the appeal from 'Says who that this is an appropriate matter for decisions to be made in that kind of forum?' cannot itself lie to mass-democratic decision making, for precisely the kind of reason just sketched. So it looks as if the inherent regress is only postponed by our long-confirmed habit of adopting a democratic criterion in 'normal' politics where it looks as if it will work roughly adequately in practice.

We can only escape this regress, in fact, by welcoming the intuition with which we are surely prompted here: that to *Says who?*, the response *Says me!* (or at any rate, *Says us!*) is the only real answer that can be given. In matters of genuine authority, political or other, *'self-appointed' is not a criticism*, but an acknowledgement of the only possible source that doesn't undermine itself. Authority cannot depend on the meeting of any criterion, but must be *assumed* and *conceded*, ongoingly, as between real individuals and groups. That is the only manner of its emergence that isn't susceptible to the regress question.

It is for exactly the same kind of reason that accusations such as 'self-appointed moralist' or 'self-appointed prophet' miss their mark. Who *else* could appoint a moralist or a prophet? (The Mayor? some committee?

election by secret ballot?) By the same token, who but *himself* could have given Gauguin his mission as a painter, or Beethoven his as a composer? – I choose clear cases of genuine mission, prime examples of exacting creative duty rather than self-inflating delusion or rationalisation. In all the central affairs of life, in fact, the only real authority arises in this way from individuals in creative action, and extends as far as it is conceded by those on whom they impinge, and then by those on whom those impinged upon impinge, and so on as far as its impetus carries – which may, in particular conjunctures, be as far as the reach of a language, an image or an idea.

But the idea of authority legitimated by self-appointment in this way only makes sense if it is from the dark self, in natural responsibility, that both claim and endorsement come. The ego-self can lay claim to no such authority over ego-selves, all of which stand on exactly the same basis as itself. True authority in any matter must be *personal* – it cannot lie in meeting any criteria that seek to impersonalise it, because the regress that we have been noting immediately raises the issue of their dependence on the *persons* applying them. But such personal authority can only come from wholeness of being, because it is only as whole organic-conscious beings that we differ radically enough for the life-relations of creative authority to arise. Ego-selves, each taking itself as fundamentally an end-in-itself ('My life is for me'), and each recognising in Kantian rationality that others must be exactly equally self-given as ends, can have no ultimate ground for exercising the living *dominance* in which authority must consist. That is why the intuition of our dark selves, the wholeness as which we can never explicitly grasp ourselves, is always available only as a power from beyond the 'I' – and why those in whom that power manifests itself as compelling insight, creative power, expressive force and practical grasp have an authority to which the abstract notional equality of egos is irrelevant.

That person or agent is *legitimately* authoritative, in any sphere, who speaks compellingly from wholeness – who channels his or her own daemon compellingly. The 'daemon' is how the Greeks thought of the dark self, and it is a metaphor that we would do well, in a general project of retrieval, to retrieve. Of course, there is always the danger of mistaking clamant ego-assertion for the daemon (in oneself or others) and a daemon may have destructive as well as creative manifestations. But that is one of the risks of life – there are no guarantees. There are no criteria of legitimation that will do for us the essential work of recognising living authority when we encounter it.

Ego-selves cannot ever lay claim to such authority, because they finish at their centredness and *have* no daemon behind the 'I'. They cannot claim to speak from more than themselves, because on their terms of being there *is* no 'more'. Thus for a community of ego-selves, all identically living in the 'for me', democratic-egalitarian head-counting is the only legitimation that makes sense. Because the ego-self has become the default self in modern conditions, and human affairs meanwhile have to be carried on, this has become our default model of legitimation, and for many ordinary purposes, as I have acknowledged, it works well enough. But in our present situation, insofar as this book has pictured it correctly – and indeed in any human situation recognised as having tragic depth – it must be a wholly inadequate source of recognition and action.

The deep issues of politics turn on the sense of human nature, of value and purpose, of the overall human predicament, which is made tacitly in the shared self-understanding of a community of people whose lives are 'politically' interwoven. In relation to these issues, that political movement or impetus is legitimately authoritative, which creatively expresses the dae-mon of its time. It does so by speaking that shared tacit self-understanding from beyond the enumerable ego-selves who are each individually mem-bers of the community – that is, from what we have earlier called the 'dark community' in its living wholeness. The political issue in retrieval is therefore also in part a matter of our re-learning how to hear authoritatively from our shared wholeness – one more aspect of the need to retrieve a habitable *humanitas*.

There are historical periods when tradition is legitimately authoritative in this way, playing the role that Burke classically assigned to it:[25] being accepted as defining the terms of ongoing political contestation because it represents an honoured intergenerational continuity standing behind the immediate desires and interests of the individual selves forming the current generation. But there are also periods when the shared self-understanding of the dark community is moving through radical transition and develop-ment. In such periods there is no criterion for the legitimacy of authority beyond the laying claim to it in ways that are found, in the event, to have been profoundly compelling. Thus the American revolutionaries were legitimately authoritative in their sphere, and even more dramatically the Jacobins in revolutionary France; they expressed the deep understanding, shared in dark commonality even by those who opposed them, that *liberty*, *equality* and *fraternity* made the real human sense of that historical moment. So too with the Bolsheviks' expression of the idea of common ownership of the means to common benefit ('from each according to his ability, to each according to his needs'). So too with the revolution in thought and purpose that established the British Welfare State, a programme going far beyond what could plausibly be held to have been endorsed in the collec-tive gasp of relief that was the Labour landslide of 1945.

That we are in an equivalently decisive period once again should be clear to anyone who has found the arguments of this book at all persuasive. As we grapple with the unprecedented *agon* of political humanity's relations to our own naturalness and to the rest of the natural world, true authority is therefore likely to lie only with a clear-sighted, existentially resilient constituency, which has seen and measured the tragedy involved in 'progress', and has acknowledged correspondingly the demands of natural responsibility. Those who can speak thus out of natural responsibility will speak from the wholeness in which they are creative nodes of the shared life from 'beyond the I' in which all participate. What such a constituency says and does will be a living outgrowth of wild humanity in its currently deepest self-understanding, and thus unimpugnable. Here is the essential licence for a vanguard social agent. This will be a movement that knows itself to represent the survival instinct and life-energy of the whole community, which unashamedly recruits as many of those with effective power as it can, which seizes initiatives as they present themselves, waits for no majoritarian permissions and makes no apology for acting in the living interests of all. It will buttress its actions where possible through the available forms of representative popular endorsement, but this will be done in order to ensure pragmatically that as much as possible of the body politic moves together in response to its leading, rather than from misconceived scruples about legitimacy. A movement operating in this spirit is *already* legitimately authoritative, and warranted in whatever use it makes of the institutions through which it must work.

At the moment the nearest thing we have to such an agency is the environmental movement, broadly so described. To play its proper role, however, it will need to become a very different kind of movement from what we are used to calling 'environmentalist', and it might be questioned whether that will still be the right word for it. Perhaps 'retrievalist' should replace it? Certainly, our prospects of retrieval will depend on the chances of such a movement's building itself up from the beginnings, in Transition and similar responses to our real situation, which we can already see emerging.

The global context: a right to life?

Finally in this outline sketch of how we might approach thinking about retrieval, we must confront and say at least something about what is undoubtedly the most difficult of its implications. This is, to put it bluntly, that retrieval as we have been considering it cannot conceivably be envisaged for a global population of seven going on nine billion human beings.

In the first place, this is because mere survival through the consequences of already-ensured climate destabilisation cannot be so envisaged. Mere survival requires a territory and ecological resources not impacted drastically in the short to medium term by climate consequences, and that condition will certainly not be met everywhere across the globe. Where it might have a chance of being met (in temperate NW Europe and Scandinavia, for example), the possibility of retrieval also depends, as we have been exploring, on the presence of a coherent community within that territory (one sharing a culture, local heritage and pattern of social arrangements that the community recognises as defining where it *belongs*). These conditions will in turn make possible levels of economic activity that may turn out to be reasonably resilient for that community, but that are increasingly unlikely to generate any distributable surplus. The clear corollary of all this is that we must think of retrieval as going on where it can while simultaneously, elsewhere in the world, the consequences of climate destabilisation are producing drought, famine, attendant social breakdown and armed conflict – as a result of which, millions will be dying.

Equally, the areas where survival and retrieval may be possible will need to be actively defended by their inhabitants, if that possibility is indeed to be realised, against growing pressure from climate refugees. These refugees are likely, in general, to be pressing in from the South upon the North – in Europe, for instance, from Mediterranean areas that are in turn being pressed upon by refugees from Africa and the Middle East. Unless areas of defensible retrieval do so defend themselves, *humanitas* will simply be lost everywhere instead of being retrieved somewhere.

But for many people this kind of prospect is just morally unacceptable, and ruled automatically out of consideration on that basis. It is at bottom a reaction of that kind, I suspect, that drives the willed optimism of people such as Lynas and Porritt. We *must* be able to go on assuring a resource base for all the predicted global population, whatever heroic technological assumptions that involves – because if we propose or even imagine a response to climate change on any other basis, it seems, we are treating parts of that population as expendable in our own interests. And by a particularly repugnant irony, we should be so treating people from areas of the world (Africa and the poorer parts of Asia in particular) whose activities had contributed least to the climate destabilisation, the consequences of which they would be suffering most drastically.

Those in the comfortable mainstream, who avail themselves of the slipperiness of sustainable development in order to go on with business very largely as usual, are assuaging that moral anxiety – insofar as they register it – in gross bad faith. But activist thinkers such as Lynas and Porritt are really driven by exposure to it, even if what they are driven into is a willed optimism of denial. Such anxiety, however, also impedes recognition of the denial that it drives, and thus inhibits a turn towards possible retrieval on the part of the very people whose vanguard role, as we have just been seeing, is going to be essential.

Nothing that we can do at the conceptual level is going to prevent this from being a terrible situation. That it is coming upon us represents a large part of the reason why our plight is tragic. But conceptual work may help us understand better the *kind* of terrible situation it will be – and in particular, the kind of value-profile that it will exhibit.

At the root of the thought that pursuing survival and retrieval locally while others starve *cannot be contemplated*, is the idea that all human beings existing on the planet have an equal right to life. Behind that idea, in turn, is something like the Kantian picture of every human being as an end-in-himself (or -herself), with its corollary that no human being can *justifiably* say of any other human being: he or she is there to serve my interests merely – and therefore, by extension, expendable if my interests so require. It follows that 'cannot be contemplated' here means more than 'cannot be contemplated without appalled horror and a determination to do all we feasibly can …' The last clause there implicitly recognises that *beyond a certain point* (although that point would have to be a very great way further down the line of serious international aid than we have ever yet got), we might have to concede that matters are out of our hands and that we must concentrate on our own prospects. But the Kantian 'cannot' is an absolute: courses of action to which it applies are morally ruled out, full stop. This thought is captured by the idea that rights are *trumps*, that rights-claims have to be met before any other considerations of moral weight can enter.[26] And a right to life seems to be the most fundamental and compelling trump card that anyone could play. If every one of nine billion people on the globe holds this card equally, just in virtue of having been born human, then policy options that don't include as a basic assumption the aim that all people should survive, and as far as possible flourish, are disqualified from the start.

But the considerations foregoing in this book should help us to see how this is a misleading picture. We don't all have an equal right to life – not because some have a greater right than others, but because the idea of a *right to life* at all can be seen, from the perspective of natural responsibility, to be illusory. That means that projecting and conditioning for a future in which we survive but millions, perhaps billions, die, is *not* simply ruled out. It must and will be morally excruciating in a variety of complex and conflicted ways – but that is a different matter.

How do our intuitions of natural responsibility prompt this recognition? The point is that 'rights' characterise relations among ego-selves. My life is for me, yours for you: that is not just the basis on which we negotiate potential conflicts under liberalism, but the basis of our moral obligations to one another, insofar as those are thought of in terms of rights and claims. Because each of us is a centre of meaning and purpose, and we can recognise that others are identically centres of their own life-purposes, there is nothing in our being as ego-centres that could justify any of us in subordinating others to his or her ends regardless of theirs. None of us could have any ground for doing this that would not also be a ground for surrendering our own purposes to those of another. We must therefore recognise as rights in others what we claim in our own right. Here comes out very clearly the grounding of this picture of morality in the way certain approaches to our relations with others *logically* undercut themselves.

But this is not enough to generate a right to life, unless my life is something from which *I* am *distinct* enough to lay an intelligible claim to it. We have seen that as an ego-self, I do indeed appear thus distinct from the material conditions of existence that constitute my living embodiment. But the intuitive awareness of my dark self in which natural responsibility consists is also an intuitive recognition that as a living whole I just *am* this embodiment, even though I cannot represent myself to my conscious self as so being. My life is something that I happen *in* or *as*, not something that happens *to me*, such that it could make sense to think of my having any claim, against anyone, that it should happen one way rather than another, or at all.

Of course, I might be said to have a right to life in the sense that I have a claim on anyone who contemplates killing me intentionally to further their own interests or purposes, that they not do so. But that does not extend to a claim on everyone that they provide me with the means to live – nor indeed do I have that claim on anyone (at random) who could so provide me. I might have it on a particular person in certain kinds of situation: if I haven't enough to eat, and you, who could genuinely help, deny me food so that you can eat more than you need, you are breaking an at least prima facie obligation of justice to me, and correspondingly ignoring a legitimate claim that I have. But this is a situation that can arise only between specific people. My claim on *you*, arising thus, that you not treat me as a means to your ends or as expendable, doesn't however carry over into a claim on *everyone*, that they support my chance-arisen autonomous existence. That would be, precisely, to have a general claim to my own life, of the kind that I have in respect of my property, my rights over which constitute claims on everyone, who- or wheresoever, that they acknowledge my ownership.

Our lurking awareness that a claim of that kind to what we too readily call 'my' life does not really make sense, is what intuitive attention to natural responsibility corroborates. As whole living beings, we cannot stand to ourselves in any such claimant relation.

Recognising this is still going to leave us, as I have said, with excruciatingly demanding moral difficulties to confront as the consequences of climate destabilisation close in. What obligations of justice *do* people with enough to eat actually have to people half-way around the globe who are starving? How do we factor into this assessment issues of differential historical responsibility for global warming? More painfully still, what moral responsibilities would the crew of a warship, or the government to whom they answer, have towards a vessel full of refugees who will not turn back from landing on a prohibited coast, when that prohibition has been imposed in defence of the carrying capacity of a retrievable community?

It may well be that such questions are not just excruciating, but strictly unanswerable. Certainly, it requires a peculiar obtuseness towards our tragic situation to assume that there is a morally right answer to be found in all such cases, if only we can apply the right framework of analysis. Of course, that does not mean that we should not devote hard thinking to making what moral sense we can. And let me re-emphasise (since I am conscious of having laid myself open to hostile misrepresentation even more on this matter than in relation to democracy) that there is no plausible moral sense to be made of our present situation that does not include *vastly* increased assistance to the innocent victims, both close-up and distant, of our 'advanced' way of life. But it does not take much imagination to suppose that these, and other dilemmas like them, may simply not be resolvable by such thinking.

To take a different and milder instance of the same thing in illustration, consider the intense marketing of personal and home care products now being pursued by international corporations such as Unilever in rural India, aided by television commercials and improved road connectivity and appealing to village dwellers in a context of improved education and increasing disposable income:

> The bare, one-room house is starved of light and furniture ... Srilatha lays out her wares on the single bed for want of a table. Satyamma's thin, arthritic fingers pick out an assortment of items: toothpaste, talc, detergent, soap, shampoo, tea bags ... 'Are you happy with the products?' I ask ... The old lady crooks her head ... She picks up a Vim dishwashing bar. 'Before,' she says emphatically, 'we washed our dishes with ash'. Enough said. Politely, she ushers us out.

It's difficult to argue with such logic … Yet, I wonder where it will stop. Are Barbie dolls an 'essential'? Or SUVs? Or shopping centres? They are to some. Are they to India's villagers? Not now. But they could be soon.[27]

Are such developments *justifiable*? The question seems impertinently absurd – and in any case, who are we, busily generating much greater *per capita* emissions from our ash-free kitchens, to ask it? Yet can it remain *un*asked, when the fully-appreciated logic of these developments involves the habitability of the Earth? (It was after all no Westerner but Mahatma Gandhi who asked: if it took Britain half the resources of the planet to achieve its prosperity, how many planets will it take a country like India?[28]) It may be that with issues like this, we have gone beyond the kind of arena in which coherent moral judgement can be achieved.

Perhaps, in the end, our human best can only be to follow the promptings of natural responsibility through the attempted retrieval towards which they guide us, doing what we can find clear moral warrant for but leaving the overall upshot to be what it will. If that is so, then seeing through the illusion of a universal right to life can make what is likely to become our necessary course at least *thinkable*, if not any more palatable. Retrieval is also about regaining the sense of natural responsibility that allows that to be a living possibility.

★ ★ ★ ★ ★

This has been, as it was always going to be, a very unsatisfactory chapter. Inevitably I have only scratched the conceptual surface of the issues for retrieval in planning, education, politics and international relations which I have discussed. And I have done so, in the last two cases particularly, with a starkness that may have seemed wilfully provocative and controversialist.

This has been partly for sheer lack of space to canvass many of the necessary subtleties and qualifications that readers may have missed. But it has also reflected a desire to exhibit clearly, in all its starkness, the difference between retrieval as I picture it, responding to our tragic situation as I have tried to understand it, and conventionally-established kinds of environmentally-informed thinking about our prospects.

If I am right, such thinking is widely vitiated by its ongoing, unacknowledged commitment to pretending. Whether or not you feel that this chapter has offered some useful tools for thinking differently, will be a measure of how right you are inclined to feel that I am.

Notes

1 The passage here is from Thomas Homer-Dixon, as quoted in Barry (2012), p. 25. See also Homer-Dixon (2006).
2 Dennett (2013).
3 Or used to use? – they probably now use some kind of chemical overkill for preference.
4 Hopkins (2008), p. 54.
5 See HM Government (2013), especially Chapter 4 ('Healthy and Resilient Communities').
6 See for instance http://incredibleediblenetwork.org.uk/about
7 HM Government (2013), p. 6.
8 Weintrobe (2013), p. 41.
9 Hopkins (2008), p. 172.
10 See (again) Foster and Gough (2005).
11 From his remarks as United States Secretary of Defence made at a press briefing on 12 February 2002: see www.youtube.com/watch?v=GiPe1OiKQuk
12 Churchill (1949) – the quotations are from pp. 88 and 246 respectively.
13 President Carter's *Address to the Nation on Proposed National Energy Policy*, 18 April 1977 – see www.youtube.com/watch?v=-tPePpMxJaA. William James used the term 'the moral equivalent of war' in a pacifist essay written for the American Association for International Conciliation, February 1910, now available e.g. at www.constitution.org/wj/meow.htm
14 Anyone familiar with William Morris' utopian romance (Morris, 1890/1970) will recognise the Transition movement as one of its lineal descendants.
15 See Scruton (2012).
16 Hopkins (2008), pp. 110–11.
17 See Collini (2012), especially Chapter 3.
18 Dilthey (1976), p. 220 – my italics in the quotation.
19 Quoted in Barry (2012), p. 104. The footnote referred to is on the same page.
20 Marx and Engels (1848/1967), p. 95 (from Chapter II: 'Proletarians and Communists').
21 Read on transition towns: http://rupertsread.blogspot.co.uk/2008/02/transition-towns-are-great-but-they.html
22 Hopkins on Read: http://webcache.googleusercontent.com/search?q=cache:http://transitionculture.org/2008/02/12/rupert-read-misses-the-point-about-transition-initiatives/&strip=1. The later quotation about voting comes from here as well.
23 Address from the Oval Office, 15 June 2010, see www.youtube.com/watch?v=Gh76 oepKFc8
24 Wolff (2006), p. 1.
25 See Burke (1790/1993).
26 See for instance Dworkin (1984).
27 Balch (2012), p. 99.
28 Quoted in Moolakkattu (2010), p. 153.

CODA: CAN WE LEARN?

We are now in a position to summarise the whole of the argument that I have presented in this book. Such an overview is worth setting out, not simply as a conventional rite of conclusion, but because, while much of this argument has been anticipated by others, the core of it is novel – and that unfamiliarity may well have interfered with its being grasped in its totality. Once it is grasped, moreover, it will be seen to lead us directly to what is probably the most important question of all.

I started from the recognition that not only those with vested interests in business as usual, nor additionally the large majority of the merely acquiescent, but even also environmental activists, are in essentially the same kind of denial about the real extent of our plight. That commonalty of denial indicates, I suggested, the power of the grip that progressivism exerts upon us. The deep-seated conviction that the point of human life, its structure of significance, is the permanent betterment of the individual and then of the social condition has become unchallengeable – and in particular, no social or political future is readily conceivable that is not driven by that assumption. This mindset is what has brought us to the position where our environmental tragedy cannot in any public discourse be seen for what it is, but has to be misrepresented as a set of problems, to which every proposed solution then reiterates the approaches and attitudes out of which the 'problems' arise.

I suggested on these grounds that the lock-hold which progressivism has over our minds must go very deep indeed. Anything taken for granted alike by people who are vandalising our only planet and by people who are striving to preserve its ecological integrity must operate more profoundly

and insidiously than any economic structure, ideology or even psychological disposition could begin to explain. I have therefore sought to locate the
source of that lock-hold in the fundamental conditions of human cognition
and action. Those are the conditions of reflexive consciousness under which
cognition is *by* a self, and the source of action is *from* a self, which we must
construct as ego over against the world presented to us in experience. But
that world is the *natural* world, the world in which as natural creatures we
are nevertheless convinced that we must wholly belong. Our conditions of
consciousness and action, that is, inherently prevent us from *simply being* the
kind of creature which, on general empirical grounds, we must understand
ourselves to be.

That sounds like a claim that human beings must be radically disabled
by their own nature. But this fissure inherent in specifically human being
has been recuperated, I have argued, for most of our species history since
we became reflexively aware, through the intuitive access to our organic-
conscious wholeness which we have maintained through living in close
active communion with the life-forces of wild nature.

In an accelerating process of positive feedback over the last three centuries, however, industrialisation has supplanted that interactivity with
nature by an increasing technological dominion over it, and the associated
accelerating urbanisation and artefactuality of our living conditions have
increasingly divorced us from that recuperative contact. Bacon's 'effecting
of all things possible'[1] for the ongoing betterment of the human condition,
massively productive of material benefit as it has been, has thus ended by
robbing us not just of an ecologically stable resource base, but of the essential conditions for making sense of our humanity – of ourselves, our lives
and our mortality.

Progressivism, the pathological project of escaping from this condition
in permanently restless forward movement, has taken its deep hold on us
(and especially in the advanced West and North, which form the crucible
for these global developments) as an instinctive, desperate recoil from that
loss of sense. The feedback process that it drives has now reached a level
of intensity at which it is seriously eroding its own human and resource
bases. It has indeed, to the best of our ability to predict these matters (if we
are honest), passed the point of no return beyond which crisis and at least
partial breakdown are inevitable.

There is no going back. We cannot for the foreseeable future recover the
kind of collective communion with the non-human wild that has sustained
us for most of the human past. Attempts to do so, I argued in connection
with 're-wilding', are essentially sentimentalising and self-delusory – that is,
only subtler forms of despair. Our only real hope is to accept what is being

forced on us by climate change, not just as a set of real consequences of what we have actually done, but as a symbolic re-irruption of the necessary and finally inescapable wild back into our lives. That must mean, into our self-understanding as well as our material life-conditions: hence the literally vital importance of recovering – and in the first place, just of making conceptual space for – what I have called *existential resilience*. Only so, I ended by claiming in the last chapter, will we be able to build the on-the-ground economic, social and political resilience required for retrieval, with even an outside chance of success.

What that summary makes painfully plain, however, is that our hopes for survival and retrieval must depend on our ability to learn from, rather than simply repeating and reinforcing, the ways in which our tragic plight has come about. *Can* we do that? Is humanity up to it?

The difficulty that we face in answering that question with any show of plausibility is that there are no previous examples to which to point, or on which to draw for hope. Other civilisations have collapsed, as Jared Diamond in particular has carefully recorded.[2] But, as he also makes clear, this has always happened through failures either of anticipation or of resolve, or through the straightforward social or technical inadequacy of their attempts to address the difficulties that confronted them. Probably no civilisation before our own, however – certainly none that Diamond examines – has addressed its crisis with 'solutions' having failure built into them, solutions designed to 'problematise' conditions that the civilisation needs to be able to think about as problems in order to think about them at all, even while it also tacitly recognises them as tragically rooted and critically threatening. At that historically unprecedented depth of practical and metaphysical entanglement, it would seem, we can only turn to works of imagination for help.

The comparatively short piece to which I myself want to turn here, and with some brief reflections on which I want to end, is E.M. Forster's frighteningly prescient story *The Machine Stops*.[3] This was first published in 1909, and was written, Forster tells us, as 'a reaction to one of the earlier heavens of H.G. Wells' – that is, to an early literary manifestation of the progressivism that is now bearing us such bitter fruit.

The story describes what Forster must have taken to be a far-future world in which most people have lost the desire and indeed the ability to live on the surface of the Earth, a large proportion of which is now dust and mud, supporting no life. Each human being now lives alone in a standard underground cell, and these cells, stacked in layered subterranean 'cities' across the globe, are all linked and powered by a vast global network, the Machine. Travel is unpopular and rarely necessary. Communication occurs by means of the 'speaking apparatus', an early form of webcam through which people conduct their only shared activity, the exchange of ideas and information.

As a consequence of such a form of life, humans are much changed physically. The main protagonist of the story, Vashti, is introduced as 'a swaddled lump of flesh – a woman, about five feet high, with a face as white as a fungus'. She is toothless and hairless, and so habituated to having all movement of herself and of surrounding objects performed mechanically for her that she 'totters' on the few occasions when she has to walk by herself. Humans of that era have also advanced (as they would themselves put it) both intellectually and emotionally. Each lives in isolation, and children (meetings for whose propagation are arranged by the relevant Committee of the Machine) are removed to the public nurseries at birth, except for those promising too athletic a physique for a wholly sedentary life, who are destroyed. Each person is in mediated contact with several thousand others, but rarely issues from his or her room to encounter another human being face to face. Time is spent in constant discussion of second-hand ideas and of the daily round of existence, interspersed by the delivery and audition of an endless succession of ten-minute lectures on a huge variety of subjects. (Vashti herself lectures on musical history, although she is shown making no music, and hearing it only mediated through the Machine.)

It is obviously possible to overemphasise the similarities between this imagined state of affairs and our own present situation – that of the advanced urban-megalopolitan societies at least, which the rest of the world strives to emulate. Forster, after all, was writing a satire intended as a warning. But it is possible to underemphasise them, too: things have moved much faster than he could have guessed, or feared. We have not yet retreated underground, nor rendered the surface of the Earth uninhabitable. But we have already made long strides in the direction of the kind of retreat from life that Forster's Machine-dwellers exemplify – towards the meaningless connectedness and the technically empowered helplessness of which they are the symbolic culmination. Much of our urban living might as well be underground, since it goes on in airless or air-conditioned, neon-lit, largely identical rooms and offices, or in shops defending themselves against the outer air with walls of wasteful heat at their entrances. We have not yet become white, physically enfeebled blobs like Vashti – many people indeed actively cultivate physical fitness in order to resist the tendencies in this direction threatened by mechanical civilisation. But the scores whom one sees running in the urban parks, with headphones clamped in their ears piping the artificial entertainment from which they seemingly cannot break free even in vigorous physical activity, do not suggest much resistance to machine-assumptions.

Even less comfortable are the similarities between interpersonal communication in the Machine-world and in ours. We do not all live alone, but historically unprecedented numbers of us now do (29 per cent of UK households were single-person at the last count[4]), a trend that has surely been greatly accelerated by the rapid development of the internet, with its ready facilitation of virtual relations and interactions without movement or engagement. (It facilitates too, as we have lately learned, a huge extension of the surveillance society, if not yet to Machine levels where every cell is monitored round the clock.) In respect of the social media which the internet has also enabled, similarities to Forster's vision are especially striking. It is not just a matter of the strong tendencies towards isolation which computerised communication encourages, nor the having scores of 'friends' compatibly with having few actual friends – with all the mental health implications, particularly for young people, of that possibility. The emergence of phenomena such as Twitter demonstrates that Forster's century-old intuition about the utterly trivialising effects of such technologies, the reduction of conversation to chatter and attention-spans to a bare minimum, was entirely well founded.

But it is the deepest similarities that are really worrying. For towards the end of his story, the Machine enters upon the slow, inevitable breakdown to which his title refers. At long last, it has become too huge and complex for any human mind fully to understand or to control it. Glitches start to appear, minor matters at first (dislocations in the piped music), initially resented but then accepted. 'And so with the mouldy artificial fruit, so with the bath water that began to stink, so with the defective rhymes that the poetry machine had taken to emit.' It emerges that the ubiquitous Mending Apparatus, universal rectifying agency hitherto of all problems with the Machine, is itself in need of mending. In response, refusal to admit what is really happening embeds itself, fuelled by rumours of new power-centres that will put everything right and do the work even more magnificently than before. All this is supported by the re-establishment of religion, in the form of a growing habit of venerating and praying to the Machine, while treating the enormous instruction manual issued to all its denizens as a holy book. But there is no help for it: in due time, the Machine indeed stops. One day, without any warning, the entire global communication system suddenly breaks down, silence floods back (many people dying instantaneously from the shock), and the Machine-world disintegrates in mass panic and general annihilation.

After what I have argued in these chapters about tragic human overreaching, systems 'too big to fail', the pervasiveness of denial, structural reliance on the problem as solution and the quasi-religion of progress, the parallels in all this with our real present situation confronting climate break-down should be devastatingly apparent. But then, recognising that, we must ask even more insistently – what is there in this imagining for hope?

So far, however, I have only recounted one side of the story, and there is another. Vashti has a son, Kuno, for whom (despite having seen him for years only through the Machine – he lives on the other side of the globe) she retains a rather un-Mechanical fondness. Kuno is a rebel. He persuades Vashti to endure the brief airship journey (Forster at least failed to anticipate aeroplanes) across half the world to his cell, where he tells her of his craving to escape the sanitised, stifling world of the Machine. He has deliberately exercised to develop his mobility and physical independence, and has found an escape route to visit the surface of the Earth, in a place that was once Wessex, with a respirator but without an egression permit. There he not only realises that he can adapt to breathing the outer air, but sees other humans living outside the Machine – evidently people who have been punished with 'extrusion', which is supposed to entail death, but who have somehow managed to survive in places on the Earth's actually not-quite-barren surface.

Kuno's brief escape is soon over – he is recaptured (ironically, by the Mending Apparatus) and moved from beneath Wessex to a cell in Vashti's own city. But he has seen what he has seen, and that becomes crucially important at the end.

For Kuno, we must suppose, cannot be unique. The inherent yearning of human beings for our natural primeval kinship with the living Earth that he represents, cannot be stifled. Not long after his escapade, we are told, respirators and egression permits are abolished – the Machine can never become so omnipotent as not to fear these profound instincts in what are now, plainly, its captives. That must be one strong sign for hope.

There are two others – even amidst the collapse of this dreadful and doomed experiment in one kind of human possibility. The first is just that it does, inevitably, collapse. However long it may last, the Machine has built into it the template for its own destruction, as the natural order of human fallibility irresistibly reasserts itself. The moral for our own experiment in ignoring natural limits is that breakdown (in our case, through the oncoming consequences of climate change) can be what saves us from our own folly and gives us another chance.

The second gleam of hope is that there is, after all, another chance. The vast, worldwide hubristic insult to human being which we have con-structed is already starting to disintegrate, like the Machine, under its own internal strains and the external pressures of the real world on which we have tried to impose it, and that disintegration will end more probably than not – again, like the Machine – in chaos and blood. But hope for a human future lies in those who have been able to see it for what it is, and who have maintained in spite of it their contact with the *essential* wild. Hope for survival and retrieval remains with the wild in us which is our responsibility to living nature, and with those who have survived 'living wild', in this sense, outside the system.

At the close of Forster's story, as their subterranean hive of networked cells is rent asunder by an airship out of control and they await their imminent deaths, its two protagonists briefly find one another and have time for a parting exchange:

> 'But Kuno, is it true? Are there still men on the surface of the earth ? Is this – tunnel, this poisoned darkness – really not the end?'
>
> He replied:
>
> 'I have seen them, spoken to them … They are hiding in the mist and the ferns until our civilization stops. Today they are the Homeless – tomorrow …'
>
> 'Oh, tomorrow – some fool will start the Machine again, tomorrow.'
> 'Never,' said Kuno, 'never. Humanity has learnt its lesson.'[5]

That is powerfully imagined. We can only go on in the faith that it could yet turn out to be true.

Notes

1 From his *New Atlantis*, in Bacon (1627/1906).
2 See Diamond (2005), especially Chapter 14 ('Why Do Some Societies Make Disastrous Decisions?').
3 Included in Forster (1954). The comment about Wells is from Forster's Introduction (p. 6 of this edition).
4 See *Families and Households, 2012*, Office for National Statistics (2012), available at www.ons.gov.uk/ons/dcp171778_284823.pdf
5 Forster (1954), p. 146.

BIBLIOGRAPHY

Bacon, F. (1612/1937) *Essays by Francis Bacon*, ed. Geoffrey Grigson, London: Oxford University Press.

—— (1627/1906) *The Advancement of Learning and New Atlantis*, intro by Thomas Case, London: Oxford University Press.

Balch, O. (2012) *India Rising: Tales from a changing nation*, London: Faber & Faber.

Barry, J. (2012) *The Politics of Actually Existing Unsustainability*, Oxford: Oxford University Press.

Batchelor, S. (1997) *Buddhism without Beliefs: A contemporary guide to awakening*, London: Bloomsbury.

Blackburn, S. (1998) *Ruling Passions*, Oxford: Oxford University Press.

Blackmore, S. (2005) *Conversations on Consciousness*, Oxford: Oxford University Press.

Blake, W. (1946) *Poetry and Prose of William Blake*, ed. G. Keynes, London: Nonesuch.

Broome, J. (2012) *Climate Matters*, New York: Norton.

Burke, E. (1790/1993) *Reflections on the Revolution in France*, ed. L.G. Mitchell, Oxford: Oxford University Press.

Carel, H. (2008) *Illness: The cry of the flesh*, Stocksfield: Acumen.

Carson, R. (1963) *Silent Spring*, London: Hamilton.

—— (1965) *The Sense of Wonder*, New York: Harper & Row.

Churchill, W. (1948) *The Second World War, Vol. 1 The Gathering Storm*, London: Cassell.

—— (1949) *The Second World War, Vol. 2 Their Finest Hour*, London: Cassell.

Cohen, S. (2001) *States of Denial: Knowing about atrocities and suffering*, Cambridge: Polity.

Collini, S. (2012) *What Are Universities For?* London: Penguin.

Daly, H. (1992) *Steady-State Economics*, London: Earthscan.

Dennett, D. (2013) *Intuition Pumps and Other Tools for Thinking*, London: Penguin.

Diamond, J. (2005) *Collapse: How societies choose to fail or succeed*, London: Penguin Books.

Dilthey, W. (1976) *Selected Writings*, trans. and ed. H.P. Rickman, Cambridge: Cambridge University Press.

Diogenes Laertius *trans.* R.D. Hicks (1925) *Lives of Eminent Philosophers*, London: Heinemann.

Dworkin, R. (1984) 'Rights as Trumps', in Waldron 1984, pp. 153–67.

Dyer, G. (2008) *Climate Wars*, Oxford: Oneworld.

Eliot, T.S. (1963) *Collected Poems 1909–1962*, London: Faber.

Forster, E.M. (1954) *Collected Short Stories*, Harmondsworth: Penguin.

Foster, J. (2008) *The Sustainability Mirage*, London: Earthscan.

Foster, J. and S. Gough (eds) (2005) *Learning, Natural Capital and Sustainable Development*, Abingdon: Routledge.

Freud, S. (1961) *The Standard Edition of the Complete Psychological Works of Sigmund Freud*, London: Hogarth Press.

Fromm, E. (1978) *To Have or to Be?* London: Cape.

Fukuyama, F. (1992) *The End of History and the Last Man*, London: Penguin.

Gardiner, S. (2011) *A Perfect Moral Storm: The ethical tragedy of climate* change, Oxford: Oxford University Press.

Georgescu-Roegen, N. (1971) *The Entropy Law and the Economic* Process, Cambridge, MA: Harvard University Press.

Goldsmith, E. *et al.* (1972) *A Blueprint for Survival*, Harmondsworth: Penguin.

Gray, J. (2002) *Straw Dogs: Thoughts on humans and other animals*, London: Granta.

—— (2004) *Heresies*, London: Granta.

Hamilton, C. (2010) *Requiem for a Species*, London: Earthscan.

Haq, G. and A. Paul (2012) *Environmentalism since 1945*, Abingdon: Routledge.

HM Government (2013) *The National Adaptation Programme: Making the country resilient to a changing climate*, London: The Stationery Office.

Hölderlin, F. (1951) *Sämtliche Werke (Band 2)*, Stuttgart: Kohlhammer.

Homer-Dixon, T. (2006) *The Upside of Down: Catastrophe, creativity and the renewal of civilisation*, London: Souvenir.

Hopkins, R. (2008) *The Transition Handbook: From oil dependency to local resilience*, Totnes: Green Books.

Hursthouse, R. (2007) 'Environmental Virtue Ethics', in R. Walker and P. Ivanhoe (eds) *Working Virtue*, Oxford: Oxford University Press, pp. 155–72.

Huxley, A. (1932) *Brave New World*, London: Chatto & Windus.

Jackson, T. (2009) *Prosperity without Growth*, London: Earthscan.

Jacobs, M. (1991) *The Green Economy*, London: Pluto Press.

Kafka, F. (1992) *Metamorphosis and Other Stories*, trans. M. Pasley, London: Penguin.

Kant, I. (1781/1999) *Critique of Pure Reason*, trans. P. Guyer and A. Wood, Cambridge: Cambridge University Press.

Keats, J. (1954) *The Letters of John Keats*, ed. F. Page, Oxford: Oxford University Press.

Kuhn, T. (1962) *The Structure of Scientific Revolutions*, Chicago, IL: University of Chicago Press.

Kumar, S. (2008) *Spiritual Compass: The three qualities of life*, Totnes: Green Books/ Finch Publishing.

La Rochefoucauld, F. (1967) *Maximes*, Paris: Editions Garniers Frères.

Larkin, P. (1988) *Collected Poems*, London: Faber.

Lasch, C. (1991) *The True and Only Heaven: Progress and its critics*, New York: Norton.

Lawrence, D.H. (1915/1997) *The Rainbow*, Oxford: Oxford University Press.

—— (1928/1960) *Lady Chatterley's Lover*, Harmondsworth: Penguin.

—— (1961) *Fantasia of the Unconscious and Psychoanalysis and the Unconscious*, London: Heinemann.

—— (1962) *The Collected Letters of D.H. Lawrence*, ed. H.T. Moore, London: Heinemann.

Leavis, F.R. (1975) *The Living Principle*, London: Chatto & Windus.

Leibniz, G.W. (1686/1934) in *Gottfried Wilhelm Leibniz: Philosophical Writings*, ed. G.H.R. Parkinson, London: Dent.

Leopold, A. (1949) *A Sand County Almanac*, Oxford: Oxford University Press.

Lomborg, B. (2001) *The Skeptical Environmentalist: Measuring the real state of the world*, Cambridge: Cambridge University Press.

Lovelock, J. (2009) *The Vanishing Face of Gaia: A final warning*, New York: Basic Books.

Lynas, M. (2007) *Six Degrees: Our future on a hotter planet*, London: Fourth Estate.

—— (2011) *The God Species*, London: Fourth Estate.

Macaulay, T.B. (1848/1906) *The History of England from the Accession of James II*, London: Dent.

McEwan, I. (2010) *Solar*, London: Jonathan Cape.

McKibben, B. (2010) *Eaarth: Making a life on a tough new planet*, New York: Times Books.

Marx, K. and F. Engels (1848/1967) *The Communist Manifesto*, trans. S. Moore, Harmondsworth: Penguin.

Metzinger, T. (2009) *The Ego Tunnel: The science of the mind and the myth of the self*, New York: Basic Books.

Mill, J.S. (1848/1994) *Principles of Political Economy*, ed. Jonathan Riley, Oxford: Oxford University Press.

—— (1874) *Nature, the Utility of Religion, and Theism*, London: Longmans, Green, Reader & Dyer.

Monbiot, G. (2013) *Feral: Searching for enchantment on the frontiers of rewilding*, London: Allen Lane.

Montaigne, M. de (1580/1965) *Essays*, trans. F. Bowman, London: Edward Arnold.

Moolakkattu, J. (2010) 'Gandhi as a human ecologist', *Journal of Human Ecology*, 29(3): 151–8.

Morris, W. (1890/1970) *News from Nowhere*, ed. J. Redford, London: Routledge & Kegan Paul.

Muir, J. (1901/1997) 'The wild parks and forest reservations of the West', in *Nature Writing*, ed. William Cronon, NY: Library of America.

Nagel, T. (1979) *Mortal Questions*, Cambridge: Cambridge University Press.

Nietzsche, F. (1906/1968) *The Will to Power*, trans. Walter Kaufmann and R.J. Hollingdale, New York: Vintage Books.

Nisbet, R. (1980) *History of the Idea of Progress*, London: Heinemann.

Norgaard, K.M. (2011) *Living in Denial: Climate Change, Emotions and Everyday Life*, Cambridge, MA: MIT Press.

Nussbaum, M. (1990) *Love's Knowledge: Essays on philosophy and literature*, New York: Oxford University Press.

Perry, J. (1979) 'The problem of the essential indexical', *Noûs*, 13(1): 3–21.

Porritt, J. (1984) *Seeing Green*, Oxford: Blackwell.

—— (2006) *Capitalism as if the World Matters*, London: Earthscan.

—— (2013) *The World We Made*, London: Phaidon.

Porter, T. (1995) *Trust in Numbers: The pursuit of objectivity in science and public life*, Princeton, NJ: Princeton University Press.

Roszak, T. (1981) *Person-planet: The creative disintegration of industrial society*, London: Granada.

Sartre, J.-P. (1943/1958) *Being and Nothingness*, trans. Hazel Barnes, London: Methuen.

Schopenhauer, A. (1813/1974) *On the Fourfold Root of the Principle of Sufficient Reason*, trans. E.F.J. Payne, La Salle, IL: Open Court.

—— (1819/1958) *The World as Will and Representation*, trans. E.F.J. Payne, New York: Dover.

—— (1841/2009) *The Two Fundamental Problems of Ethics*, trans. D. Cartwright and E. Erdmann, Oxford: Oxford University Press.

Schumacher, E.F. (1974) *Small is Beautiful: A study of economics as if people mattered*, London: Sphere.

Scruton, R. (2012) *Green Philosophy: How to think seriously about the planet*, London: Atlantic Books.

Skidelsky, R. and E. Skidelsky (2012) *How Much is Enough?* New York: Other Press.

Snow, C. (1959) *The Two Cultures and the Scientific Revolution*, Cambridge: Cambridge University Press.

Soper, K. (1995) *What is Nature?* Oxford: Blackwell.

Soper, K. and F. Trentmann (eds) (2008) *Citizenship and Consumption*, Basingstoke: Palgrave Macmillan.

Sprat, T. (1667/1959) *History of the Royal Society*, edited with critical apparatus by Jackson I. Cope and Harold Whitmore Jones, London: Routledge & Paul.

Strawson, P. (1974) *Freedom and Resentment and Other Essays*, London: Methuen.

Tawney, R. (1926) *Religion and the Rise of Capitalism*, London: Murray.

Taylor, P. (1986) *Respect for Nature: A theory of environmental ethics*, Princeton, NJ: Princeton University Press.

Tolstoy, L. (1921) *A Confession and What I Believe*, trans. A. Maude, London: Oxford University Press.

Urry, J. (2005) 'The Complexity Turn', *Theory, Culture and Society*, 22(5): 1–14.

Waldron, J. (ed.) (1984) *Theories of Rights*, Oxford: Oxford University Press.

Weber, M. (1905/1992) *The Protestant Ethic and the Spirit of Capitalism*, trans. Talcott Parsons, London: Routledge.

Wegner, D. (2002) *The Illusion of Conscious Will*, Cambridge, MA: MIT Press.

Weintrobe, S. (ed.) (2013) *Engaging with Climate Change: Psychoanalytic and interdisciplinary perspectives*, Hove: Routledge.

Williams, B. (1978) *Descartes: The project of pure enquiry*, Harmondsworth: Penguin.

Wilson, E. (1984) *Biophilia*, Cambridge, MA: Harvard University Press.

Wittgenstein, L. (1961) *Tractatus Logico-Philosophicus*, trans. D.F. Pears and B. McGuinness, London: Routledge & Kegan Paul.

Wolff, J. (2006) *An Introduction to Political Philosophy*, Oxford: Oxford University Press.

Yeats, W.B. (1919/1950) *Collected Poems*, London: Macmillan.

INDEX